R for Earth and Sustainability Data Analytics

Maps show where.
Time series show how.
Data analytics helps us understand why.

R for Earth and Sustainability Data Analytics

A Beginner's Guide to Earth Data Analytics with GIS, Mapping, Time Series, and Machine Learning in R

Antonios E. Marsellos, Ph.D.
Katerina G. Tsakiri, Ph.D.

Earth Data Analytics Press

2026

ISBN: 979-8-9965635-0-0 (Paperback)
ISBN: 979-8-9965635-1-7 (eBook)
ISBN: 979-8-9965635-2-4 (Hardcover)

Library of Congress Control Number: 2026915118

PUBLISHER Earth Data Analytics Press
Queens, New York

AUTHORS
Antonios E. Marsellos, Ph.D.
Department of Geology, Environment, and Sustainability
Hofstra University

Katerina G. Tsakiri, Ph.D.
Department of Information Systems, Analytics and Supply Chain Management
Rider University

COVER DESIGN
Cover design by the authors.

First Edition
First published in 2026

Contact
For questions, feedback, or educational inquiries, please contact the authors.

Disclosure: The authors are co-founders and owners of ResForMe LLC (resforme.com), a company that provides research mentorship services to high school students. As owners of the company, the authors may benefit financially from references to ResForMe contained in this book. Any discussion of ResForMe is intended solely for educational and informational purposes and does not imply endorsement by the authors' affiliated universities or institutions.

Brief Contents

Table of contents

Preface

Why Coding Matters in Earth and Sustainability Data Analytics

As more fields rely on data, the ability to analyze, interpret, and communicate information effectively is essential across disciplines. When we study data tied to places (geospatial analysis), environmental science, or changes over time (time series modeling), there are two main ways to work with data: graphical software and coding. Both have value. The difference is how far they can take you.

Coding lets you do what graphical interfaces cannot.

Graphical software is genuinely useful. It is fast, visual, and practical. Imagine being in the field collecting data on your phone or tablet. You need to see a map, click a few buttons, and make a quick decision. A graphical interface is perfect for that moment. It is also the easiest way to introduce GIS or time series analysis to beginners.

But graphical tools are built for convenience. They are not built for full control. Menus change, buttons move, and workflows evolve. A process you learned today may look different in a few months. Repeating the exact same analysis later can become difficult.

Coding works differently. It is stable. Every step is written, visible, and repeatable. If you run the same code tomorrow, next year, or on another computer, you get the same result. Instead of depending on where a button is located, you are defining the entire process yourself.

From Clicking to Thinking

Transparency is one of the biggest advantages of coding. You can see every step, and changing one line immediately shows its effect.

Think of a simple example. You process satellite data and realize that one parameter needs adjustment. In a graphical tool, you may need to repeat the entire workflow step by step, which can take hours. In coding, you change one value and run the script again. The result updates in seconds.

Coding changes how you work. Instead of repeating steps, you refine ideas. In graphical workflows, improvement often means repeating the process; in coding workflows, it happens instantly. This speed encourages exploration: you can try more ideas, test more hypotheses, and reach better conclusions.

Coding Inside vs Outside Software

Some graphical tools offer scripting options. SPSS, GIS platforms, and other analytical software allow users to write code within their systems. At first glance, this seems like the best of both worlds. You get the convenience of a graphical interface with the power of coding, and you can automate tasks while still using familiar tools. In practice, it often creates new challenges.

These scripting environments are usually limited, harder to debug, and less flexible. They depend on proprietary systems and smaller user communities. When something goes wrong, finding help can be difficult. Being outside of a specific software environment, using languages like R or Python, provides more freedom. You can use any free add-on or tool that the worldwide community of users has shared (these free add-ons are called libraries), without being tied to a specific platform's limitations. When you encounter an issue, you can search for solutions across a vast ecosystem of users and resources. This makes learning and troubleshooting much easier.

R or Python?

Millions of users work with these languages. There are tutorials, forums, examples, and shared solutions for almost every problem. If you search for an issue, chances are someone has already solved it.

Even generative AI tools perform better with these languages. They have been trained on large amounts of public code, which makes their suggestions more accurate and more useful.

Choosing a language is not only about how you type commands (its syntax). It is about choosing a community and a support system you can rely on for years.

Why R

Tools like Excel, SPSS, and Tableau are useful in specific contexts. Excel is simple and familiar. SPSS provides structured statistical workflows. Tableau creates appealing visualizations. However, they often separate the workflow. You may clean data in one tool, analyze it in another, and visualize it somewhere else.

R brings everything together. You can import data, clean it, analyze it, model it, and visualize it in one place. You can also integrate geospatial analysis directly into the same workflow.

Historically, GIS began with coding. Early systems such as GRASS GIS relied on command line tools in UNIX environments. Later, graphical systems like ArcView made GIS more accessible. Today, tools like ArcGIS Pro and QGIS are widely used.

R reconnects GIS with its computational roots while keeping modern flexibility. It allows you to design your entire workflow from start to finish.

Learning One Language Helps You Learn Others

A common question is whether to learn R or Python. R is strong in statistics, visualization, and research workflows. Python is broader and widely used in software development and machine learning.

Both are powerful. More importantly, both teach you how to think computationally: how to break a problem into clear, repeatable steps a computer can follow. The real goal is not choosing a language but learning how to design solutions. Learning one language does not limit you. It gives you a foundation to learn others. The skills you develop in R will transfer to Python, and vice versa.

The New Reality: Coding with Generative AI

A major shift has occurred in recent years. Students no longer need to understand every line of code to begin producing meaningful results. With generative AI, a user can write or generate advanced scripts, run them, and immediately see outputs. This creates an important contrast. In graphical software, even simple workflows require time. Each step must be performed manually. There is no way to compress dozens of clicks into a few seconds.

In coding, a full workflow can run instantly. You can generate a script, execute it, observe the result, and refine it in a continuous loop. What once took hours or days can now take minutes.

More importantly, coding in environments such as R allows users to immediately access advanced analytical tools. Within seconds, a student can move from raw data to visualization, geospatial mapping, and time series analysis, and even apply advanced prediction methods (with names like neural networks and random forests) that you will meet, step by step, later in the book. These are techniques that are widely used in industry and often associated with well-paid analytical work.

Seeing meaningful results early changes motivation. Instead of spending weeks learning software mechanics before producing useful output, students can explore the entire data analysis workflow from the beginning. That immediate feedback encourages curiosity, confidence, and a stronger desire to learn.

Understanding still matters. Over time, users should learn what their code does and how each step works. However, the entry point is no longer a barrier. Students can begin by exploring results and gradually build deeper understanding as they refine their workflows. It reflects how many professionals work: they start with a question, investigate it with code, and learn the details along the way.

This is a fundamental shift in how we teach data analytics. It allows us to focus on creativity, exploration, and problem-solving from the start rather than getting stuck in mechanics. It is also an unusually accessible time to begin research and data analysis because advanced computational tools are more widely available than ever before.

A Balanced Perspective

Graphical tools and coding do not compete. They complement each other. A graphical interface is like using a prepared meal kit. It is fast and helpful when you need something ready. Coding is like being a chef. You control the ingredients, the process, and the final presentation. Each approach has its place. But only coding allows unlimited creativity.

Final Thought

A graphical interface helps you complete a task. Coding allows you to design the entire workflow. It gives you the power to create solutions, not just use them. In Earth and Sustainability Data Analytics, this means you can ask new questions, explore environmental and societal systems, and develop new analytical workflows for real-world problems. You are not limited by what a software company has built; you can build your own tools. This book is an invitation to move from using tools to creating solutions.

About This Book

R as a Tool

Earth data are everywhere. From satellites and weather stations to groundwater sensors, climate records, environmental monitoring networks, and GPS systems, the Earth continuously generates information about how our planet is changing. The challenge is not access to data; it is knowing how to understand it.

This book was written to help students take their first steps in that process. Every figure, table, map, analysis, and example in this book was created using R. By building the entire book through code, we follow the same step-by-step process you will learn, one that anyone can run again and get the same result (this is called a reproducible workflow).

Rather than focusing on programming for its own sake, this book uses R as a tool to explore real Earth and environmental problems. The goal is not to memorize commands, but to develop a way of thinking: how to organize data, recognize patterns, ask meaningful questions, and interpret results in a scientific context.

The structure of the book reflects how real analysis works. We begin with the basics of R and gradually move toward working with real-world datasets. Along the way, we explore key ideas in data handling, spatial analysis, time series, and modeling. Each chapter builds on the previous one, using consistent examples so that concepts feel connected rather than isolated.

A central idea throughout the book is that simple methods, when used carefully, are often more powerful than complex techniques that are not fully understood. For this reason, the emphasis is on clarity, interpretation, and workflow. Every step is explained in a way that helps you understand not only what the code does, but why it matters.

Think Like an Earth Data Analyst

This book assumes no prior experience with R or programming. R is a free program you install on your computer to work with data, and we will show you exactly how to set it up in Chapter 1. If you are new to coding, you may feel uncertain at first. That is normal. The goal is not to learn everything at once, but to become comfortable reading code, running scripts, and gradually building confidence.

By the end of the book, you should be able to:

- read and understand basic R scripts,
- work with real geoscience datasets,
- visualize spatial and time-based patterns,
- build and interpret simple models,
- and approach new problems with a structured analytical workflow.

Most importantly, you should begin to think like a data analyst working across Earth and sustainability systems: asking questions, testing ideas, and interpreting results carefully and responsibly.

This book is not about getting perfect results. It is about learning how to ask better questions. Every question is worth asking; some answers simply need more thinking. Taking the time to understand a problem is often more important than trying to solve it immediately. Our goal is to give you real answers when possible, and when they are not, to show you how to approach them.

Acknowledgments

We would like to express our gratitude to the many students and colleagues who have shaped the way we think about teaching data analytics in Earth science.

Through years of teaching geostatistics and spatial analysis across universities in Europe and the United States, we have learned a simple but important lesson: students do not struggle with complexity; they struggle with abstraction. When concepts remain theoretical, they are easily forgotten. When they become hands-on, they become meaningful. This book was written with that philosophy in mind.

One recurring observation has been the limitation of "click-and-go" software. While such tools are easy to learn initially, they often become outdated as interfaces change and workflows evolve. Tutorials that rely on sequences of clicks rarely remain useful beyond a short period of time. In contrast, programming provides a more stable and transferable way of thinking. Functions, logic, and structure persist over time, even as tools evolve. For this reason, we strongly believe that coding is not just a technical skill, but a long-term investment in how students approach problem-solving in Earth and Sustainability Data Analytics.

We would like to thank Gail Bennington, adjunct faculty member and former director of the Hofstra University Summer Science Research Program (HUSSRP), for providing the opportunity to mentor high school students in research. Working with students who had no prior coding experience, yet successfully developed projects and presented their work at scientific conferences and competitions, reinforced a central belief behind this book: that with the right approach, anyone can learn to think analytically, work with data, and contribute meaningfully to research, regardless of their starting point.

We would like to acknowledge Charalambos Petrocheilos, whose introduction to R programming fundamentally influenced the way we approach data analysis. This experience reinforced a guiding principle that shaped this book: *think twice, code once.*

Finally, we are deeply grateful to our families for their patience and understanding during the many moments when this book required time beyond our professional responsibilities. Their support made this work possible.

How This Book Progresses

This book is designed as a practical introduction rather than a comprehensive reference manual. The emphasis is on understanding workflows, interpreting patterns, and developing analytical thinking through real Earth and sustainability applications. This book is organized as a gradual progression from understanding data to interpreting Earth systems analytically.

- Chapter 1 → learning R
- Chapter 2 → preparing reliable data
- Chapter 3 → understanding spatial structure
- Chapter 4 → understanding temporal structure
- Chapter 5 → understanding relationships, prediction, and an introduction to machine learning
- Chapter 6 → applying Earth and sustainability data analytics

Final Integration:

- Chapter 7 → understanding spatial and temporal global change and sustainability

Each chapter builds on the previous one, using consistent datasets and examples to create a connected learning experience. The goal is not to cover every possible method, but to develop a way of thinking that can be applied to new problems and datasets in the future. As you move through the book, the focus gradually shifts. In the early chapters, you may find yourself asking, "What does this function do?" Later, the more important question becomes, "Does this result make scientific sense?" Learning to evaluate results critically is one of the most important skills in Earth and Sustainability Data Analytics.

Each chapter follows a familiar learning rhythm, but the purpose is not repetition. The repeated structure is designed to help you stay oriented while the scientific reasoning becomes progressively more advanced. The book is also designed with different learning levels in mind. Beginners can successfully complete only the Part A sections in the earlier chapters and still build a strong foundation in Earth and Sustainability Data Analytics. Parts B and C gradually introduce more advanced workflows, spatial analysis, modeling concepts, and real-world applications.

Throughout this book, do not try to memorize every command. Focus on the questions the data are helping us answer. The code is only the means; scientific interpretation is the goal. By the end of the book, you should have a solid foundation in how to approach Earth and sustainability data analytically, with the tools and mindset to explore new questions on your own.

The Companion Website

Every example in this book is built from real, runnable R code, and all of it is free to readers. The companion website gathers the code, the datasets, and more in one place:

https://marsellos.github.io/rgeo

There you can:

- **Get the complete code and datasets.** Request the chapter-by-chapter R code and the datasets that go with it, so you can run every example yourself and adapt it to your own work.
- **Read about each dataset.** Every dataset comes with its source, terms of use, and a plain-language guide to its columns.
- **Find updates and corrections.** Errata, new material, and a one-page quick-start cheat sheet.

Instructors can also request a free **Instructor Resource Package**: lecture slide decks for every chapter, a teaching manual, lab activities, an assessment bank, and grading rubrics. See the *For Instructors* page on the companion website.

Part I

Foundations

Chapter 1

Introduction to R

Getting Started with RStudio and Basic R Concepts

1.1 Introduction

This chapter gives you a first look at the main ideas you will use in R. You do not need to remember everything right away. The goal is to help you feel comfortable opening RStudio, running code, and reading simple R scripts.

We will return to these ideas later in the book. You will see them repeatedly in different contexts, and they will become second nature with practice.

This chapter is organized into three parts:

- Part A introduces core concepts
- Part B introduces packages and libraries
- Part C provides optional topics for learners who want to explore further

In later chapters, you will use these same ideas to work with real Earth and sustainability data, including maps (space), time series (time), environmental systems, and human impacts.

> **i** What you will be able to do
>
> By the end of this chapter, you will be able to:
>
> - open RStudio and run your first lines of code in a script with Ctrl + Enter
> - store values in objects with the arrow `<-` and combine values into a vector with `c()`
> - use built-in functions (ready-made tools) such as `mean()`, `summary()`, and `str()` to explore your data
> - organize data in a data frame (a table of rows and columns)

- draw a simple plot with `plot()` and a map of a region with the maps package

You are not expected to memorize these. This chapter is a gentle first look, and you will practice them throughout the book.

1.2 Part A: Core Concepts

1.2.1 A Note for Windows and Mac Users

This book was written to work on both Windows and macOS. Nearly all code examples are identical across operating systems.

When keyboard shortcuts are shown, Windows users should use **Ctrl**, while Mac users should use **Command**. For example:

- **Ctrl + Enter** (Windows)
- **Command + Enter** (Mac)

Other differences between operating systems are minor and will not affect most examples in this book.

1.2.2 Before You Begin: Installing R and RStudio

Before opening RStudio, you need to install two free programs in this order:

Install **R** from https://www.r-project.org, then click "Download R" and choose your operating system. Install **RStudio** from https://posit.co, then click "Download RStudio Desktop (Free)".

Always open RStudio, not R directly. RStudio is the environment you will work in; R runs behind the scenes.

You can type the code in this book yourself as you go, which is the best way to learn, or get the complete, ready-to-run code for every example from the companion website (https://marsellos.github.io/rgeo).

1.2.3 Getting Oriented in RStudio

RStudio panes

RStudio has different sections, but you only need to focus on two at the beginning:

- **Script** (top left) → where you write your code
- **Console** (bottom left) → where you see results

Getting Started: Your First Setup

> **i** Create a new script
>
> Press **Ctrl + N** to create a new script in the top-left section of RStudio.

Think of it like this: the script is your notebook (you write here), while the console is your calculator (you see answers here). There are more panes, but these two are the most important when you start.

You will use the script to write and save your code, and the console to see the results of running that code. For reference (don't worry about memorizing this), RStudio is usually organized into four main areas:

- **Source pane (Script area)**: where you write and save your code
- **Console**: where R shows results after you run code
- **Environment / History**: shows what you created and what commands you previously ran in R
- **Files / Plots / Packages / Help**: where you view files, graphics, installed packages, and documentation

Script versus console

Run code with Ctrl + Enter

For this course, always write your code in a script and run it with **Ctrl + Enter**. The console is useful for testing, but the script is where your work should be. Code typed in the **console** runs, but disappears when you close R. Think of the console like scratch paper: you can test things quickly, but they are not saved. A script is like your notebook because you can save your work and return to it later.

You can run a line by placing the cursor anywhere on that line; there is no need to keep the cursor at the beginning or end of each line. You can also select multiple lines and run them together.

> ⚠ Why you should use a script
>
> Writing code only in the console means you lose your work when you close R. Always use a script for anything you want to keep.

Before continuing, it helps to know a few basic actions in RStudio such as how to create, run a script, and ask for help.

Creating and saving a script

To create a new script:

- Click **File → New File → R Script**
- Save it with **Ctrl + S**

Saving your script helps you keep your work and return to it later.

Writing comments

Sometimes you want to write notes for yourself inside your code. These notes are called comments. You can write notes in your code using the # symbol, which starts a comment. Everything to the right of # on the same line is ignored by R. Comments are not run by R. They help explain what your code is doing. For example:

```
# The hash symbol (#) starts a comment.
# R ignores comments when it runs code.
x <- 10  # store the value 10 in the object x
x
```

```
[1] 10
```

The arrow `<-` means:

"put this value into this object."

When you run these lines with **Ctrl + Enter**, R calculates the answers and shows the results in the Console. Running the object name `x` by itself tells R:

"Show me what is stored inside `x`."

You may also see `=` used for assignment in some R code (for example, `x = 10`). Both work, but `<-` is the standard style in R and is what this book uses throughout.

> **!** How to run code in R
>
> R does nothing until you tell it to run code. Use **Ctrl + Enter** (Windows) or **Command + Enter** (Mac) to run code from your script.
>
> You can run code by:
>
> - placing your cursor anywhere on a line and pressing **Ctrl + Enter**
> - selecting multiple lines and pressing **Ctrl + Enter**
> - placing your cursor anywhere inside a multi-line command and pressing **Ctrl + Enter**
>
> You do **not** need to copy code into the Console.

Getting help

At the console, if you type:

```
?mean
```

R will open the documentation (help) page for `mean()`, a built-in tool that averages numbers, in the Help pane (bottom right). Most help pages include a description, the inputs the tool accepts, and worked examples near the bottom. (We explain functions and their inputs in detail later.)

1.2.4 Your First Commands

Basic arithmetic

When you open RStudio for the first time, focus on two areas:

- Write your code in the Script pane and run it
- See the results in the Console

A typical workflow is to write code in the script, run it with **Ctrl + Enter**, and view the results in the Console or Plots pane. If you get an error in the console, read it carefully to understand what went wrong. Errors are normal when learning programming.

```r
# Run by holding ctrl and pressing enter
# on each line
5 + 3
```

```
[1] 8
```

```r
10 - 4
```

```
[1] 6
```

```r
6 * 2
```

```
[1] 12
```

```r
8 / 4
```

```
[1] 2
```

```r
1:10
```

```
 [1]  1  2  3  4  5  6  7  8  9 10
```

We explain the : (which makes a sequence of numbers) and the [1] marker in the output just below.

Spaces

- R ignores extra spaces in most situations, although consistent spacing makes code easier to read.

In R, the : symbol creates a sequence of numbers.

So:

```
1:10
```

means:

1, 2, 3, 4, ..., 10

This is a quick way to create consecutive values.

The `[1]` is added automatically by R. It shows the position of the first value in the output. You do not need to type it yourself. It becomes useful later when working with larger results.

> **!** When R gets "stuck"
>
> If you run an incomplete command such as:
>
> ```
> 5 +
> ```
>
> The R console will show a `+` and wait for you to finish the command. To stop this and return to the normal prompt (the `>` symbol where R waits for your next command), click the console and press **`Esc`** on your keyboard. This happens often when a parenthesis, quote, or operation is not completed.

1.2.5 What Is an Object?

An object is like a labeled box where you store a value. Objects help us save information so we can use it again later without retyping it. Almost everything in R is stored inside objects.

Store a value

You can create an object by assigning a value to it. For example, `x <- 10` means "store 10 in a box called x".

This saves a value so you can use it later:

```
x <- 10
```

Think of it like putting a label on a box:

- `x` is the label
- `10` is what's inside the box

Storing a value in an object is done with the symbol `<-`. You type `<-` by pressing the less-than `<` key followed by the hyphen `-` key.

> **i Why use objects?**
>
> Imagine you calculate a temperature value once and want to use it many times later. Instead of retyping the number every time, you can store it in an object and reuse it whenever you need it.

Create your first graph

Let's visualize numbers by creating a graph using two objects, `day` and `rainfall_mm`. We will create two short lists of numbers (these are called vectors, explained in the next section) and plot them against each other.

```
# Create two vectors
# of equal length
day <- 1:5
rainfall_mm <- c(0, 12, 5, 0, 8)
```

These code lines create five day numbers (1 to 5) and the rainfall in millimetres on each day, and store them in objects called `day` and `rainfall_mm`.

```
# Create a simple plot of day and rainfall
# The graph appears in the Plots pane (bottom right)
plot(day, rainfall_mm)
```

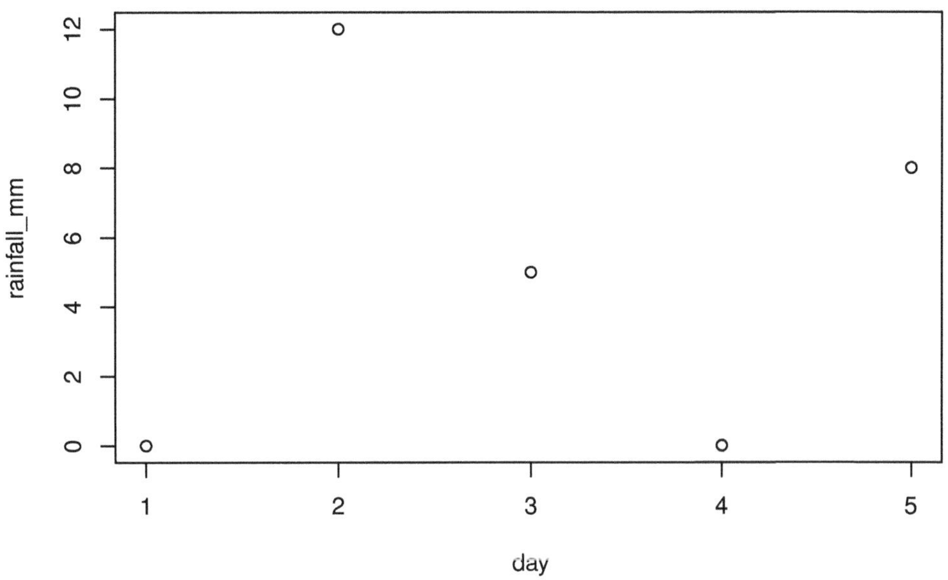

Figure 1.1: A simple scatter plot of daily rainfall (mm) over five days.

Because we gave R two vectors to plot:

- `day` becomes the horizontal axis
- `rainfall_mm` becomes the vertical axis

The rainfall changes from day to day, so the points do not form a straight line. Each dot represents one pair of values, such as `(1,0)`, `(2,12)`, `(3,5)`, and so on. The pair `(2,12)` means day=2 and rainfall=12, which places a dot above the second day at a height of 12. On days with no rain the value is 0, so the dot sits on the bottom axis.

Seeing data as a picture helps you notice patterns quickly. The rainfall rises and falls across the five days, with some days showing no rain at all (Figure 1.1). Even small datasets can reveal important structure when visualized. In geoscience, plots are one of the main ways we explore data. The `plot()` function used here comes built into R, so you can start making graphs straight away. More specialized graphing packages also exist, such as `ggplot2`, which we use later for the case studies in Chapter 6. For now, `plot()` is all you need: it is simple to write and can still produce clear, informative, and detailed graphs.

1.2.6 Vectors

A vector is a group of values stored together. Usually, all values inside a vector should be similar, such as all numbers or all text. The `c()` command (short for "combine") joins several values into one vector.

```r
temperature <- c(15.2, 16.1, 16.8, 17.4)
plot(temperature)
```

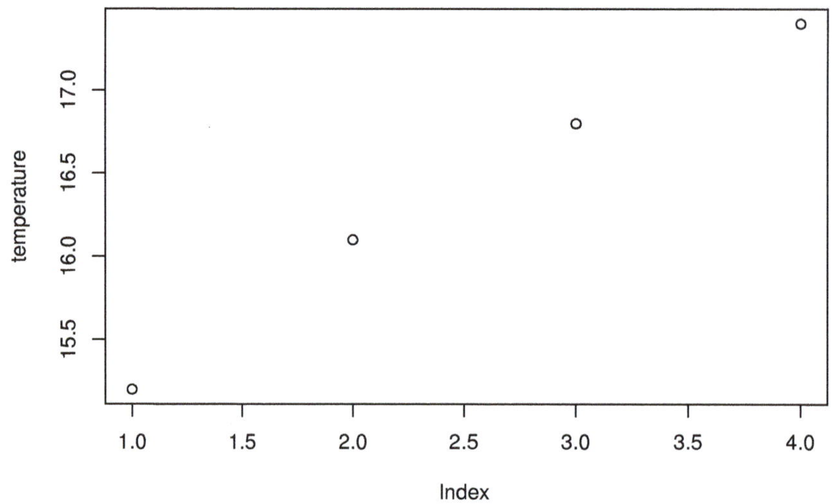

Figure 1.2: A simple plot of the four values in the temperature vector, shown against their position (index).

Vectors are commonly used in real environmental and sensor datasets because they store groups of related values together. In this example, the `temperature` vector contains four numeric values. You can think of a vector as a list of numbers that belong together. When you create a vector, all values must be of the same type (e.g., all numbers or all text).

> **i Plotting a single vector**
>
> Because we only gave R one vector, R automatically uses the positions of the values:
> 1, 2, 3, 4
> for the horizontal axis.
> So this plot shows:
>
> - position on the x-axis
> - temperature value on the y-axis

i Common mistake: mixing numbers and text

A vector usually works best when it contains similar kinds of values.
For example:

```
c(1, 2, 3)        # all numbers
c("A", "B")       # all text
```

Mixing numbers and text together can lead to unexpected results.
For example:

```
c(5, "number")
```

If you ran this, the `5` would be shown in quotes as `"5"`, meaning R now treats it as text.
You can think of it like this:

- `5` is a real number that R can calculate with.
- `"5"` is more like a picture or label of the number 5.

It is similar to the difference between:

- a real burger you can eat
- and a picture of a burger

The picture looks like a burger, but you cannot eat it.
In the same way, `"5"` looks like a number, but R treats it as text instead of something to calculate with. This can cause problems later if you try to perform calculations with that vector.
Sometimes numbers should NOT be treated as numbers.
For example:

```
zipcode <- "11549"
```

A ZIP code looks numeric, but we usually do not calculate with it.
We would not add ZIP codes, multiply ZIP codes, or calculate averages of ZIP codes. A ZIP code is really a label, not a quantity. The same is true for:

- phone numbers, groundwater monitoring station IDs
- borehole IDs, earthquake event IDs
- satellite image IDs, sample numbers
- GPS station names, geological unit codes, and many other labels

Even though some of these contain numbers, they are often **labels rather than quantities**.
Always check that your vectors contain the type of data you expect.

An object can contain one value or many elements. Besides numbers and text, objects can also store vectors, data frames, lists, and arrays (different ways of organizing data that we introduce later in the chapter).

```
name <- "Earth" # This is a text object (character)
number <- 25    # This is a numeric object
flag <- TRUE    # This is a logical object (holds only TRUE or FALSE)
```

TRUE and FALSE are special values that represent yes/no or true/false situations.

Use `str()` to quickly check what an object contains. Sometimes you forget what kind of data an object contains. The `str()` function is like asking R:

"What is inside this object?"

Sometimes R gives errors because the data type is not what you expected.

For example:

- text instead of numbers,
- or TRUE/FALSE values instead of text.

```
str(name)
```

```
 chr "Earth"
```

```
str(number)
```

```
 num 25
```

```
str(flag)
```

```
 logi TRUE
```

In this output, `chr` means character (text), `num` means number, and `logi` means logical (a TRUE or FALSE value).

The `length()` function tells us how many values are in the vector. This is useful when working with larger datasets because you often need to know how many values or observations you have.

```
length(temperature)
```

```
[1] 4
```

Before continuing, ask yourself

- What does the vector store?
- How many values does it contain?
- Which value is at position 1?

Indexing in vectors

You can pick specific values from a vector. The first value is at position 1, the second at position 2, and so on. The brackets `[ ]` are used to select values from an object.

Brackets `[ ]` are used to pull values out of an object. You can think of them like asking:

"Which position do I want to look at?"

2:4 means the values 2, 3, and 4.

```
temperature[1]   # first value
```

```
[1] 15.2
```

```
temperature[2:4] # positions 2 to 4
```

```
[1] 16.1 16.8 17.4
```

This means: "give me the first temperature value." Think of it like positions in a list (e.g. 1 = first item, 2 = second item). The second line says: "Give me values starting at position 2 and ending at position 4."

Notice the difference:

- `( )` are usually used with functions
- `[ ]` are used to access positions inside objects

Quick checkpoint

At this point, make sure you understand (you do not need to memorize) the following:

- how to store values in an object,
- how to create a vector with `c()`,
- how to run code with **Ctrl + Enter**,
- and how to access values using brackets `[ ]`.

These are some of the most important foundations in R.

Combining values into a vector

We already used `c()` earlier to combine values into a vector. A vector is a group of values stored together in one object. You will use vectors often because they are one of the simplest ways to store data in R.

By the way, `c()` and `plot()` are examples of functions: small ready-made tools that do something for you (we cover functions properly in the next section). For example, `plot()` creates graphs and other functions perform calculations. A function starts with its name, followed by parentheses `()`. R keeps reading the function until it finds the closing parenthesis `)`.

This is why multi-line functions still work correctly. You can run a multi-line function by placing your cursor anywhere within the function and pressing **Ctrl + Enter**.

```
values <- c(2, # you can run this function
            4, # from any line within it
            6,
            8)
# This line displays the object while
# the previous lines create the vector
values
```

```
[1] 2 4 6 8
```

Did you notice that the `c()` function can be written across multiple lines? R keeps reading until it finds the closing parenthesis `)`. This allows you to organize your code in a way that is easier to read.

> ⚠ Avoid using reserved names
>
> Avoid naming objects as **`data`**, **`list`**, **`mean`**, or **`plot`**. These are existing R functions and can cause unexpected errors. Use names like: data_temp, list_values, or mean_value.

Try this yourself:

```
# Try this yourself
rain_mm <- c(5, 10, 15)
snow_mm <- c(2, 4, 6)

# Calculate total precipitation
# by adding the vectors
total_precip <- rain_mm + snow_mm
total_precip
```

```
[1]  7 14 21
```

Notice how we used descriptive names like rain_mm instead of just `a` or `x`. This is a great habit to build early. It makes your code much easier to read and reminds you of the units you are working with.

When we add them, R performs the operation element by element to calculate the total for each event. In other words, R adds the values one position at a time:

- first with first → $5 + 2 = 7$
- second with second → $10 + 4 = 14$
- third with third → $15 + 6 = 21$

Simple summary

These functions help you quickly understand your data. More advanced methods for summarizing data will be covered in later chapters, but these are good starting points for exploring vectors. The `summary()` function gives a quick overview of the data, including, among other values, the smallest value, the middle value, the average, and the largest value.

```r
mean(values)
```

```
[1] 5
```

```r
summary(values)
```

```
   Min. 1st Qu.  Median    Mean 3rd Qu.    Max. 
    2.0     3.5     5.0     5.0     6.5     8.0 
```

The `1st Qu.` and `3rd Qu.` columns are called quartiles: `1st Qu.` is the value a quarter of the way through the sorted data, and `3rd Qu.` is the value three quarters of the way through. You do not need them for now.

> **i Common beginner mistakes**
>
> - Forgetting to run the code (nothing happens until you press **Ctrl + Enter**)
> - Forgetting to finish a function (for example, a missing closing parenthesis) or leaving a command incomplete (for example, `5 +`) can cause errors or unexpected behavior.
> - Running code only in the Console instead of saving it in a script (you lose your work when you close R)
> - Overwriting objects accidentally by reusing the same name

See how a previous value is lost:

```
x <- 10
x <- 5   # (10 replaced by 5)
```

- Using names that already belong to existing functions, such as `mean`, `plot`, or `data`

```
mean <- 10 # Do not run this; it is shown only to explain the mistake.
# It replaces the mean() function with the number 10,
# so restart R to restore the function.
```

- Forgetting parentheses in functions:

```
mean          # wrong
mean(values)  # correct
```

1.2.7 Functions

Using Functions

A function is a tool that does something for you. Do not worry about understanding every symbol yet.

You can **read this from the inside-out** (start with the part deepest in the parentheses, then work outward):

- `c(...)` first creates a vector
- `mean(...)` then calculates the average of that vector

```
# mean function with a vector of numbers as argument
  mean(c(1, 2, 3, 4, 5))
```

```
[1] 3
```

This function returns a single value based on the numbers you provide. You give the function input, and it returns output. The parentheses `()` are where you place the information you want the function to use. For example, `sqrt()` finds a square root, and `round()` rounds a number. The information (input) you give to a function is called an argument.

Functions are ready-made tools in R that take your input and produce a useful result.

Functions let you:

- calculate values
- create plots
- work with data
- and automate tasks

For example:

- `plot()` creates a graph
- `sqrt()` calculates a square root
- `round()` rounds a number to a specified number of decimal places

```
sqrt(16) # square root function
```

```
[1] 4
```

```
log(10)  # logarithm function
```

```
[1] 2.302585
```

```
round(3.14159, 2) # round to 2 decimal places
```

```
[1] 3.14
```

Creating your own Functions

A function has a name, parentheses (not brackets), and sometimes arguments. Let's create a simple function that adds two numbers together. This is just an example to show how to define a function in R. The function takes two arguments, `a` and `b`, and returns their sum.

```
add_two_numbers <- function(a, b) {
  a + b
}
```

And, let's call the `add_two_numbers` function with specific values to see how it works:

```
add_two_numbers(5, 10) # This will return 15
```

```
[1] 15
```

We can create a function to perform a specific task. For example, a function to convert Celsius to Fahrenheit. A function is defined with the word `function`, followed by its inputs (arguments) in parentheses and, in curly braces `{}`, the instructions the function carries out (this part is called the body). A function usually takes an input, performs an operation, and returns a result. By default, R shows the last line as the result. For example:

```
celsius_to_fahrenheit <- function(celsius) {
  (celsius * 9/5) + 32
}
# And you can call the function with a specific value:
celsius_to_fahrenheit(25) # converts 25°C to Fahrenheit, returns 77
```

```
[1] 77
```

Common mistakes when writing functions

- Not using curly braces `{}` to define the body of the function, which can lead to syntax errors
- Using the same name for the function and its argument, which can cause confusion and errors. (e.g. `celsius_to_fahrenheit <- function(celsius_to_fahrenheit) { ... })` is not recommended

Before moving on

Make sure you understand these core ideas before continuing:

- how to create an object
- how to run code with **Ctrl + Enter**
- how to create a vector with `c()`
- how to use brackets `[ ]` to access values
- and how functions use parentheses `( )`

These ideas will appear repeatedly throughout the rest of the chapter.

Parentheses vs. Brackets

Parentheses `()` and brackets `[]` have different jobs in R.

- Parentheses `()` are used with functions. They tell R to perform an action.

- Brackets [] are used to pull values out of objects. They tell R which positions you want to access.

For example:

- mean(values) calls the mean() function
- values[1] accesses the first value in the vector

1.2.8 Data Frames

A data frame is a table made of rows and columns. You will use data frames often in R because many real datasets look like spreadsheets with rows and columns.

```
df <- data.frame(                # create a data frame
  station = c("A", "B", "C"), # station names column
  temp = c(15.2, 16.4, 14.8), # temperature values column
  rain = c(2.1, 0.0, 5.4)     # rainfall values column
)
```

A very common function to explore a data frame is head(), which shows the first few rows and all columns.

```
head(df)
```

```
  station temp rain
1       A 15.2  2.1
2       B 16.4  0.0
3       C 14.8  5.4
```

The names() function shows the column names.

```
names(df)
```

```
[1] "station" "temp"    "rain"
```

Use str() to quickly see what something contains. It shows what kind of values each column contains, such as numbers or text.

```
str(df)
```

```
'data.frame':	3 obs. of  3 variables:
 $ station: chr  "A" "B" "C"
 $ temp   : num  15.2 16.4 14.8
 $ rain   : num  2.1 0 5.4
```

Here `3 obs. of 3 variables` means 3 rows and 3 columns, and each `$ name:` line is one column with its data type (the trailing `...` just means more values follow). You do not need to understand all of this yet. We will use it again when working with real data. These functions help you quickly understand the data content before working with it. Let's access a column in the data frame using a name:

```r
df$temp
```

```
[1] 15.2 16.4 14.8
```

The `$` symbol means: "Look inside this table and give me this column."

> **Tab**: one of your most useful tools in R
>
> By typing `df$` and pressing `Tab`, you can see all the column names available in the data frame. Pressing `Tab` again can autocomplete the column name instead of typing it manually. This is a helpful way to avoid typos and remember the structure of your data. You can also access a column by position (e.g. `df[ , 2]` for the second column), but using names is clearer and less error-prone.

Indexing in data frames

You can access a row by position in your data frame by using brackets [ ] (not parentheses):

```r
df[1, ]
```

```
  station temp rain
1       A 15.2  2.1
```

Inside the brackets: - the first position is the row - the second position is the column

Just like in vectors, the first row is at position 1, the second row is at position 2, and so on. The comma `,` separates rows and columns. So `df[1, ]` means:

"give me the first row and all columns".

Also, you can access a specific value by row and column position of your data frame:

```r
df[1, 2]
```

```
[1] 15.2
```

This means:

- row 1
- column 2

> **Data frames are like spreadsheets**
>
> Each column of a data frame can have a different type of data (e.g. one column can be numeric, another can be text). This makes data frames very flexible for storing tabular data. However, all values within a single column must be of the same type (e.g. all numbers or all text). Also, all columns must have the same number of rows. This is a limitation that lists and arrays do not have.

So far we have worked with simple objects. Some real-world datasets are too complex to fit into a simple spreadsheet-style table. Many Earth datasets contain measurements that vary across location, depth, or time. Lists and arrays help store this type of data.

> **You already know more than you think**
>
> At this point, you already know how to:
>
> - create objects,
> - store vectors,
> - run functions,
> - create simple plots,
> - access values by position,
> - and work with basic tables in R.
>
> These are the same core ideas used in real data analysis workflows.
> You are not expected to memorize every command. The goal is to become comfortable reading code, testing ideas, and understanding how data are organized.
> Everything from this point forward will build on these same foundations.

1.3 Part B: Packages and Libraries

R includes many built-in functions, but it can do much more through **packages**. A package is a group of extra functions you can add to R. Packages can also include datasets and tools that extend what R can do.

A simple way to think about it is:

- **R base** gives you the core tools
- **packages** give you extra toolboxes
- **functions** are the tools inside those toolboxes

At this stage, you do not need packages for every task. Most examples in this chapter use only base R functions. This is intentional, so you can first understand the core ideas: objects, functions, vectors, data frames, arrays, and loops. As you begin working with real datasets, you will start using packages. Packages add new tools to R for specific tasks. They are essential for working with real Earth and sustainability data. Many packages are created by other users and researchers. For example, the `maps` package allows us to draw geographic boundaries, such as countries and states. In later chapters, we will use more advanced packages to work with real-world data, including spatial and time series datasets.

You need to install a package only once with `install.packages("your_package")`, but you need to load it with `library(your_package)` every time you start a new R session.

```
# Install only once if needed.
# Uncomment this line the first time you use the package.
# install.packages("maps")

# Load the maps package
# Installed packages must be loaded each time you start a new R session
library(maps)
```

> ℹ Common mistakes
>
> It is easy to forget to load a package with `library()` before using functions from that package. Even if a package is already installed, R does not automatically load it when RStudio starts. Packages remain installed on your computer, but they must be loaded into the current R session each time you open RStudio or start a new session.
>
> Another common mistake is confusing `install.packages()` with `library()`:
>
> - `install.packages("package")` installs the package onto your computer and usually only needs to be done once.
> - `library(package)` loads the package into the current R session and must be run whenever you start a new R session and want to use that package.

> Also remember that `install.packages()` uses quotes around the package name, while `library()` does not.

Accessing all functions in a package

You can also call a function directly from a package when needed. This is helpful when you want to be very clear about where a function comes from. The double colon `::` means "from this package, use this function": `maps::map()` reads as "use the `map()` function from the `maps` package," even without loading the whole package first. This matters later when two packages share a function name and you want to specify which one to use.

You can also explore functions inside a package by typing the package name followed by `::` and then pressing **Tab**.

For example:

```
maps::
```

Pressing `Tab` will show available functions inside the maps package. This is useful when you do not remember the exact function name or want to explore what a package contains.

The next example draws U.S. state boundaries by calling a function directly from the maps package (Figure 1.3).

```
# Example of calling a package function directly
maps::map("state")
title(main = "Map of the United States")
```

Figure 1.3: Map of the United States generated using the maps package.

The **maps** package includes several built-in map datasets, such as `state`, `county`, `world`, `world2`, and `usa`. You can see the full list with `help(package = "maps")` or `library(help = "maps")`, which open the package documentation in the Help pane (bottom right).

- "state" → U.S. states
- "world" → countries around the world

If you want to plot a country such as Greece, use the built-in world map and specify the country name with the `regions` argument:

```
# Region names must be written in lowercase and in English.
map("world", regions = "greece", fill = TRUE, col = "lightblue")
```

Figure 1.4: Map of Greece generated using the maps package.

You can also plot the entire world map with `map("world")` to see all countries. The `regions` argument allows you to specify which country or region to display (Figure 1.4). The `fill = TRUE` argument fills the area of the specified region with color, and `col = "lightblue"` sets the fill color. The map is simplified and does not include labels. It is not meant for precise identification, but for basic visualization. You can customize the appearance of the map by changing these arguments to include more details (north arrow, scale, legend, etc.).

Or, you can zoom in on a specific region, such as New York State (Figure 1.5).

```r
map("state", regions = "new york")
title(main = "New York State")
```

Figure 1.5: Map of New York State generated using the maps package.

This example shows how we can select a region by name. In later chapters, we will work with spatial data where regions are selected based on their properties and geometry.

> **i** Region names must match the dataset
>
> Region names in the **maps** package must be written in lowercase and must match the dataset naming. For example, "New Jersey" should be written as "new jersey", as shown in Figure 1.6. If you use a name that does not exist in the dataset, you will get an error.

To check more state names use:

```r
map("state", plot = FALSE)$names
```

We set `plot = FALSE` because we want the state names without drawing a map. The `$names` part extracts the state names from the result returned by `map()`. This prints a long character vector of valid region names, beginning `[1] "alabama" "arizona" ...`, and you can use these names to plot specific states.

```r
map("state", regions = "new jersey", fill = TRUE, col = "lightblue")
title(main = "New Jersey State")
```

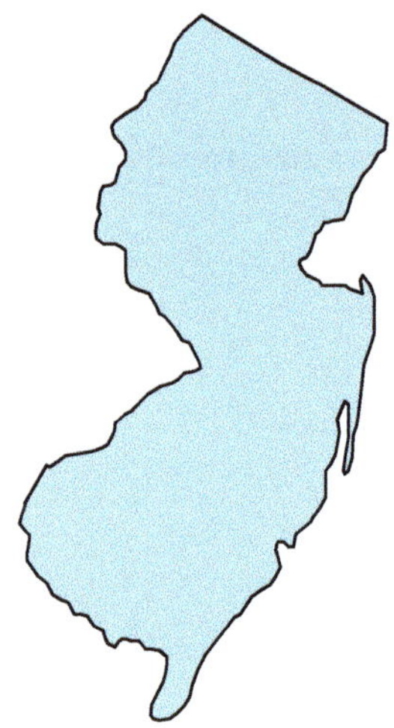

Figure 1.6: Map of New Jersey generated using the maps package.

1.3.1 A preview: interactive maps in R

Geoscience often involves working with maps. R can also create interactive, web-based maps with only a few lines of code.

> **i** A preview of what R can do
>
> The goal is not to memorize these commands. The goal is to see that R can create interactive maps and to understand that maps are another way to explore data step by step.

```
# Install only once if needed.
# Uncomment this line the first time you use the package.
# install.packages("leaflet")
library(leaflet)

m <- leaflet()   # create an empty map object
m <- addTiles(m) # add background map
m <- setView(    # set the initial view of the map
  m,
  lng = -73.6, # longitude
```

```
  lat = 40.8,  # latitude
  zoom = 8     # initial zoom level
)
m              # display the map
```

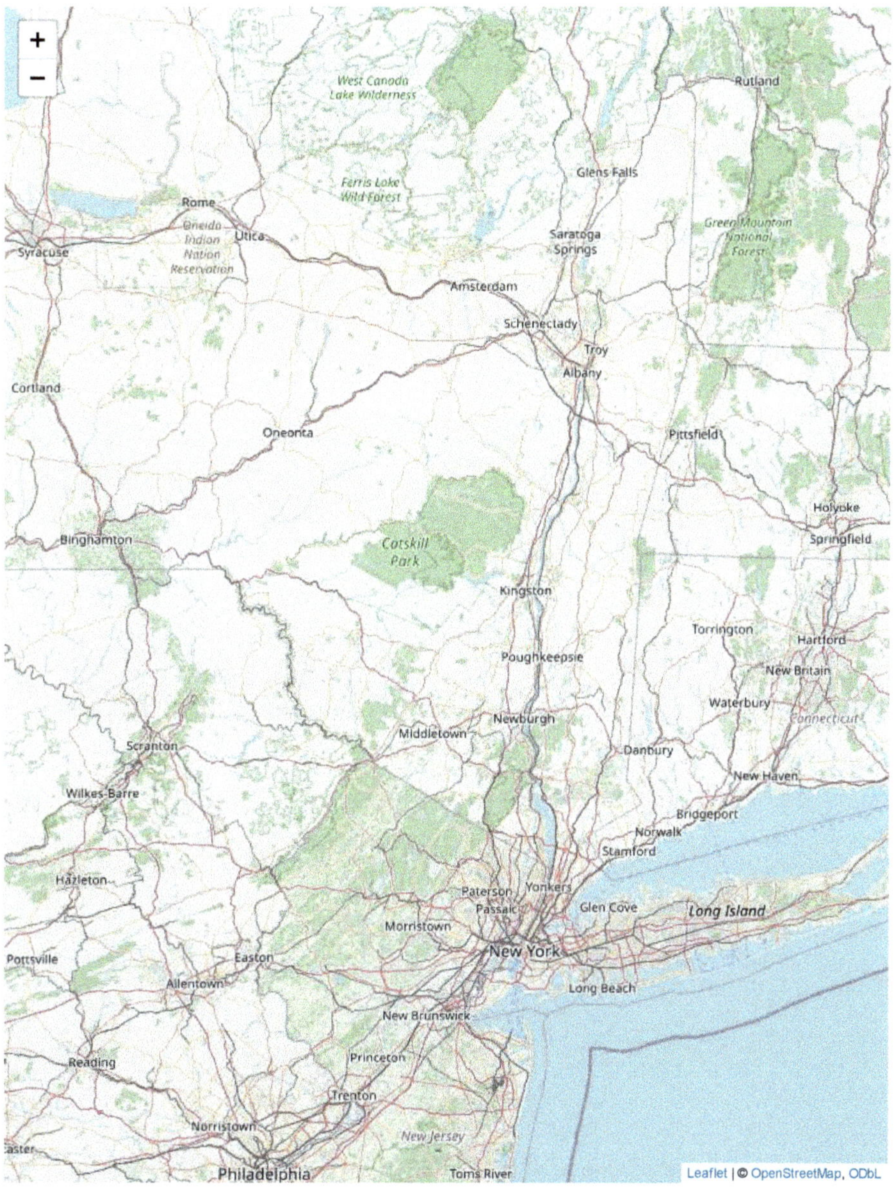

Figure 1.7: Interactive map demonstrating how R can create zoomable, web-based maps for exploring geographic information.

> **i Longitude first, latitude second**
>
> `lng` means longitude and `lat` means latitude. Together they specify the location where the map initially opens. If they are accidentally switched, the map may open in the ocean or in a completely different location.

When you run this code in RStudio, the map appears in the Viewer pane and can be zoomed and panned interactively. In this printed version of the book, it appears as a static screenshot. Interactive maps are widely used in research websites, dashboards, sustainability projects, and science communication. We will not focus on interactive mapping in this book, but it is useful to know that R can create web-based maps with only a few lines of code.

In Chapter 3, we will return to mapping and begin working with spatial data more directly.

1.4 Part C: Beyond the Basics (Optional)

1.4.1 Preview: Lists and Arrays

Some data do not fit into simple tables. In these cases, we use lists and arrays to organize them. This section is optional, but it will help you understand more complex Earth datasets later.

Lists and arrays become essential when working with real Earth and sustainability datasets. You do not need to master lists and arrays yet. For now, just know that:

- a **list** can hold different kinds of objects
- an **array** can hold values in more than two dimensions

We will return to these ideas later when they become useful.

Lists and Arrays vs Data frames

Data frames organize data in rows and columns, which works well for tabular information like station measurements. But many Earth datasets include an extra dimension. For example, temperature can change with ocean location (`latitude` and `longitude`) and `depth`. Another example is `ice thickness`, which can vary across `location` and `time`. Satellite data are often stored as grids (x,y) that repeat over time. These types of data cannot be easily stored in a simple table. Instead, we use lists and arrays to represent them.

1.4.2 Lists

A list is a container that can hold different kinds of data, such as numbers, text, vectors, or data frames. For now, just think of a list as a box that can hold different items. This is useful for storing complex data structures that do not fit neatly into a table. For example, you might have a list that contains a vector of numbers, a vector of text labels, and a data frame with measurements. Each element of the list can be accessed by its name or position.

Indexing in lists

```
my_list <- list(         # an object with different types of data
  numbers = c(1, 2, 3), # three elements vector
  label = c("model_A", "model_B"),     # two text elements vector
  table = df            # a data frame element
)
my_list
```

```
$numbers
[1] 1 2 3
```

```
$label
[1] "model_A" "model_B"

$table
  station temp rain
1       A 15.2  2.1
2       B 16.4  0.0
3       C 14.8  5.4
```

To read this output, each `$name` heading marks one item in the list: `$numbers`, `$label`, and `$table`. Under `$table` you see the whole data frame printed out.

You can access parts of a list by name, using `$` followed by the element name:

```
my_list$numbers # access the numbers vector
```

```
[1] 1 2 3
```

```
my_list$label   # access the character (label) vector
```

```
[1] "model_A" "model_B"
```

You can also access them by position. For lists we use double brackets `[[ ]]`, which pull out the actual item stored at that position:

```
my_list[[1]] # access the first element (numbers vector)
```

```
[1] 1 2 3
```

```
my_list[[2]] # access the second element (label vector)
```

```
[1] "model_A" "model_B"
```

Access the second row of the data frame stored in the list (station "B" and its temperature and rainfall values):

```
# first [[3]] pulls out the data frame, then [2,] takes its second row
my_list[[3]][2,]
```

```
  station temp rain
2       B 16.4    0
```

Lists are very flexible and can be used to store complex data structures, such as the output of a model that includes multiple variables, parameters, and results. They are also useful for organizing data that have different types or lengths, such as a list of station measurements where each station has a different number of observations.

The next examples introduce ideas that are useful later in the book, especially for spatial and time-based Earth data. Focus on the main idea: some datasets contain multiple layers that change across space and time. Lists and arrays help us organize and work with these complex datasets.

1.4.3 Arrays

The next examples are more advanced than the earlier sections. Do not worry about memorizing the syntax.

An array is like a stack of tables, maps, or images.

Each table can represent:

- a different day
- a different depth
- or a different satellite image

Arrays store data in more than two dimensions. For example, an array can store satellite images collected over many days. You can think of an array like a stack of images.

- rows run from top to bottom
- columns run from left to right across the table
- layers are separate images stacked together

Here `1:24` makes the numbers 1 through 24, exactly enough to fill the array, since $3 \times 4 \times 2 = 24$.

```
arr <- array(1:24, dim = c(3, 4, 2))
arr
```

```
, , 1

     [,1] [,2] [,3] [,4]
[1,]    1    4    7   10
[2,]    2    5    8   11
[3,]    3    6    9   12

, , 2

     [,1] [,2] [,3] [,4]
[1,]   13   16   19   22
[2,]   14   17   20   23
```

```
[3,]   15   18   21   24
```

In this output, the `, , 1` header marks layer 1 and the `, , 2` header marks layer 2. Each layer is one 3×4 table.

The dimensions `c(3, 4, 2)` mean:

- 3 rows
- 4 columns
- 2 layers

So this array contains two separate 3×4 tables stacked together.

> **i How the array is stored**
>
> R fills arrays column by column rather than row by row.
> For example, the values 1, 2, and 3 are placed in the first column before R moves to the second column.

Visualizing array layers

To access a value, use:

`arr[row, column, layer]`

Each layer of the array can be visualized as an image.

```
arr[1, 1, 1]  # row 1, column 1, layer 1
arr[1, 1, 2]  # same position, but in layer 2
```

The printed output above shows the two layers separately:

- `, , 1` represents the first layer (the first 3×4 table)
- `, , 2` represents the second layer (the second 3×4 table)

Each layer contains one 3×4 grid of values.

If you leave something empty before a comma, R understands that you are not selecting a specific row or column.

The empty spaces before the commas mean:

- "use all rows"
- "use all columns"

So this command says:

"show all rows and columns from layer 1."

```
# the image() function draws values as a colored grid; this shows layer 1
image(arr[ , , 1])
```

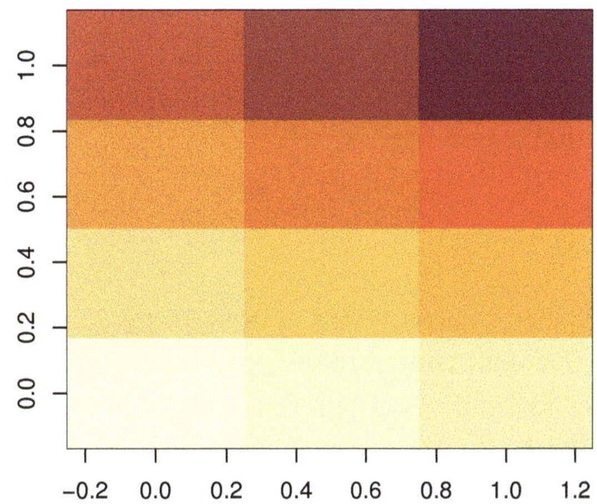

Figure 1.8: First layer of a 3D array visualized as an image. Light colors represent lower values and dark colors represent higher values.

The colored grid does not line up cell-for-cell with the printed table below, because `image()` rotates the values when it draws them.

```
arr[ , , 1] # print the first layer as numbers
```

```
     [,1] [,2] [,3] [,4]
[1,]    1    4    7   10
[2,]    2    5    8   11
[3,]    3    6    9   12
```

The image shows how values are distributed across the grid. The printed values give exact numbers, while Figure 1.8 provides a visual summary of their spatial distribution.

Each cell in the image represents one value in the array. The color shows the size of the value:

- lighter colors = smaller values
- darker colors = larger values

This helps you quickly notice differences across the layer.

Comparing the printed layer with the image helps you understand how values map to colors.

> ⚠ How the array is displayed
>
> When R prints an array in the console, you read it like a table or spreadsheet. However, when you display it with `image()`, R treats the rows and columns as positions in space rather than as a spreadsheet.
> As a result, the image may not look exactly like the printed table. Do not worry if the visual appears rotated or flipped compared with the console output. You did not make a mistake. This happens because `image()` draws rows from bottom to top (like a geographic map), while the console prints rows from top to bottom (like a spreadsheet). The values are exactly the same; only the orientation differs.

Subtracting images

You can also perform calculations between layers. For example, subtracting one layer from another can show how values change between time steps or conditions. This helps show how data change over time, such as calculating anomalies or differences between time steps. The resulting image shows where values increased or decreased between the two layers, which can represent change over time.

```
# subtract the first layer from
# the second layer and visualize the result
image(arr[ , , 2] - arr[ , , 1])
```

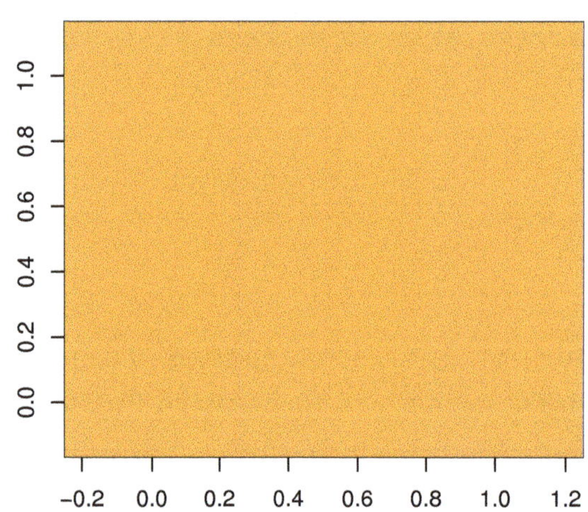

Figure 1.9: Difference between the second and first layers of the array. The uniform color indicates that every cell has the same value (12).

All values are equal (12), which results in a uniform image (Figure 1.9). The image appears as a single color because every cell has the same difference. This happens because the second layer contains values 13 to 24, while the first layer contains values 1 to 12, so each corresponding position differs by exactly 12.

Working with Multiple Layers

So far, we visualized each layer manually. But what if we had many layers (for example, 100 time steps)?

Instead of writing the same code repeatedly, we can use a for loop to automate the process. This allows us to visualize all layers in a stack without having to write separate code for each one. This is especially useful when working with large datasets, such as time series of satellite images, where you might have dozens or hundreds of layers to visualize.

1.4.4 Loops

A loop repeats the same action several times automatically.

Think of it like telling R: "Do this again and again."

Loops are useful when the same operation must be done for many values.

Imagine you had 100 images instead of 2. Writing the same plotting command 100 times would be slow and tiring. A loop lets R repeat the same task automatically.

Everything inside the curly braces `{}` is repeated.

```
for (i in 1:5) {
  print(i)
}
```

```
[1] 1
[1] 2
[1] 3
[1] 4
[1] 5
```

In this example, the object `i` takes the values 1, 2, 3, 4, and 5.

You can think of `i` as a number that changes automatically during the loop. The loop runs the code inside the curly braces `{}` for each value of `i`. So it will print the numbers 1 through 5, one at a time. Loops will be useful later when working with time series and spatial data.

A loop repeats the same action for different values. Let's use a loop to visualize both layers of our array without writing separate code for each layer:

```
# add a title to each plot
# indicating the layer number (1 or 2)
for (i in 1:2) {
  image(arr[ , , i])
  # paste() joins the word "Layer" with the number i
  # to make the title text
  title(main = paste("Layer", i))
}
```

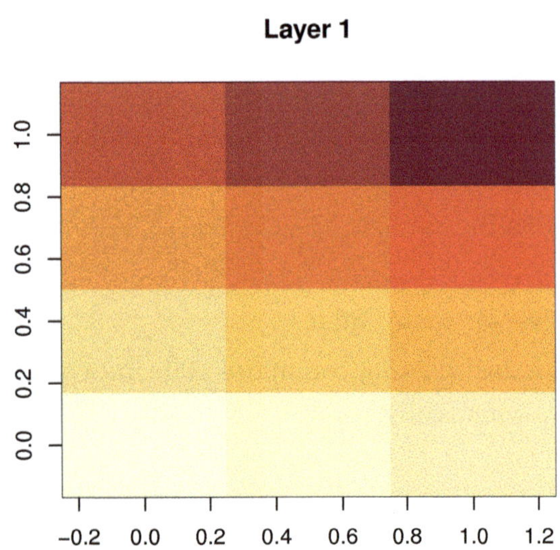

(a) Array layer 1.

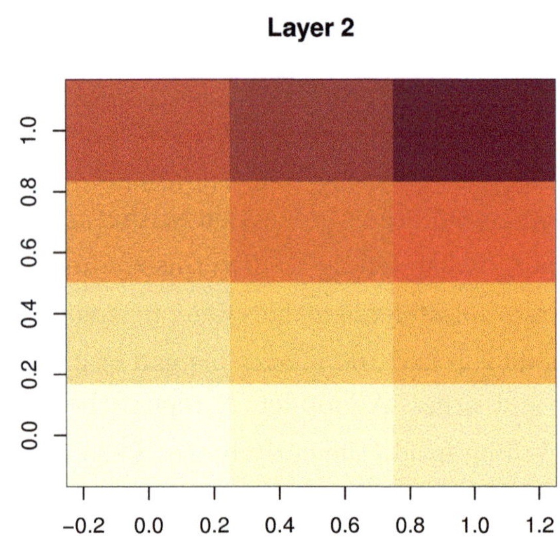

(b) Array layer 2.

Figure 1.10: The two array layers drawn by the loop.

Arrays and loops are useful when working with data that change across space and time, such as ice thickness or temperature grids. The loop runs through each layer of the array, that is, `i` takes the values 1 and 2. For each value of `i`, we plot the corresponding layer. This avoids repeating the same code multiple times.

These ideas will become important when working with real datasets, such as ice thickness changing over space and time.

> **i Curly braces define the loop**
>
> A common mistake is forgetting the curly braces { }. These show which lines belong to the loop.

> In RStudio, enabling *Rainbow Parentheses* in Global Options can make matching parentheses, brackets, and braces easier to see.

1.4.5 A Volcano Temperature Application

Now let's use a small example to connect the ideas of arrays, indexing, and plotting.

We will use the built-in `volcano` dataset as a grid of values. For this example, we will treat those values as temperatures measured across a volcanic area.

Day 1 shows the original temperature field. Day 2 shows the same field with a localized warm area in the northern part of the map. We will store these two days in a three-dimensional array, where the first two dimensions represent space and the third dimension represents time.

A 3D array is like a stack of maps. Each layer in the stack represents one day.

Here we modify a small square in the northern part of the grid to simulate localized heating, like a fumarole (a hot vent on a volcano). This creates a clear contrast with the surrounding area, making the change easier to visualize. The values `48:50` select a small square region in the northern part of the grid. This shows how one part of the data can change while the rest stays the same, which is common in Earth datasets.

```r
# Create a 3D array sized to match the volcano grid
# (rows, columns, and 2 time layers).
# NA marks an "empty"/missing value (we fill it in below).
# nrow() and ncol() count the rows and columns of volcano.
temp <- array(NA, c(nrow(volcano), ncol(volcano), 2))

# Day 1: baseline temperature
temp[,,1] <- volcano

# Day 2: localized heating in the northern part of the map
temp[,,2] <- volcano

# Add heat to the volcano area on day 2
# Simulate localized warming
# like a fumarole (hot volcanic vent)
temp[48:50, 48:50, 2] <- temp[48:50, 48:50, 2] + 50
```

What this code does

- `temp` stores the full dataset.
- The first layer is Day 1.
- The second layer is Day 2.
- The line with `48:50, 48:50, 2` changes only a small area in the northern part of the map on the second day.

Think of it like warming only a small area in the northern part of the map while leaving the rest unchanged. Many Earth datasets change across both space and time, such as temperature, elevation, or satellite observations. The next example shows how to visualize those changes side by side.

```
# par() changes global plot settings
# and mfrow = c(1, 2) arranges two plots
# side by side in one row and two columns
# mar = c(bottom, left, top, right): tight margins,
# only top kept for the Day 1 / Day 2 titles
par(mfrow = c(1, 2), mar = c(1, 1, 3, 1))
# Day 1
# col = terrain.colors(20) picks 20 map-like colors;
# axes = FALSE hides the axis numbers
image(temp[,,1], col = terrain.colors(20), axes = FALSE, main = "Day 1")
# add contour lines showing locations
# with the same temperature (called isotherms)
contour(temp[,,1], add = TRUE)
# Day 2
image(temp[,,2], col = terrain.colors(20), axes = FALSE, main = "Day 2")
contour(temp[,,2], add = TRUE)
# Restore the default plotting layout for future plots
par(mfrow = c(1, 1))
```

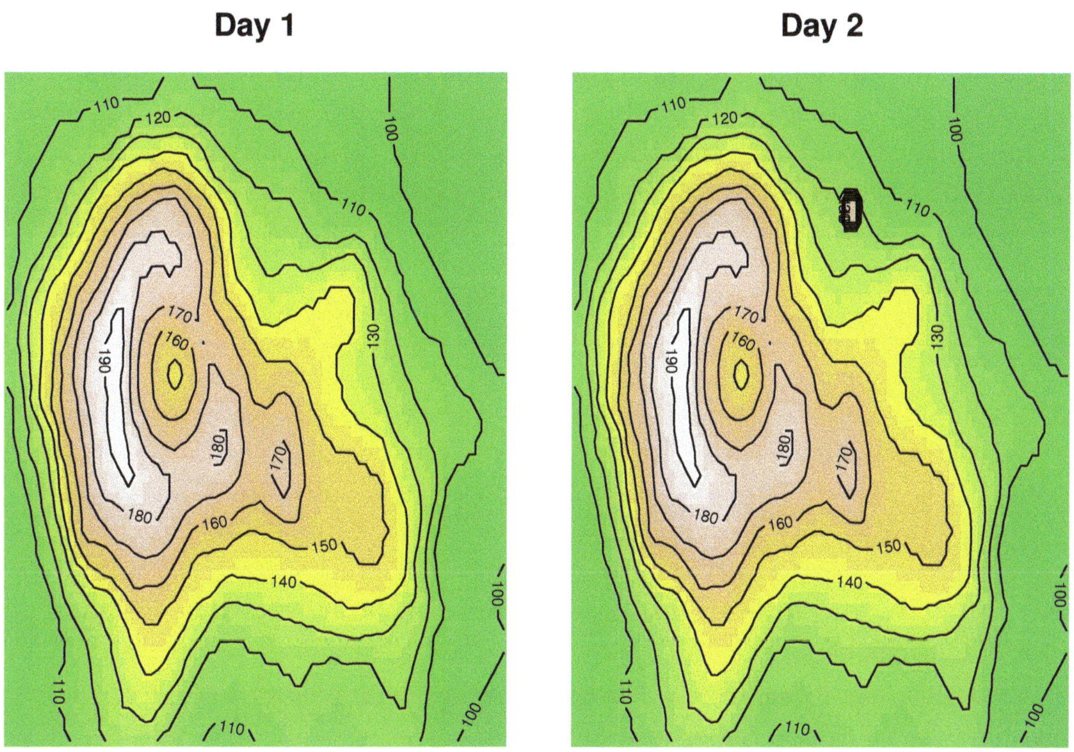

Figure 1.11: Simulated temperature over a volcano for two consecutive days. The second day shows increased temperature in a localized area of the northern part of the map, illustrating how spatial data can change over time.

Contour lines connect areas with similar values. In this example, they connect locations with similar temperatures, much like elevation contours connect locations with similar elevations on a topographic map. Every point along the same contour line has the same temperature.

The two days side by side

In this example, we treat the `volcano` dataset as a temperature field. Each value represents the temperature at a specific location on the surface.

In the three-dimensional array, the first layer corresponds to Day 1, and the second layer corresponds to Day 2.

On Day 2, we increase the temperature in a small area of the northern part of the map to simulate localized heating. This shows a common pattern in Earth data: one area can change over time while nearby areas remain similar.

> **i** Why visualization matters
>
> Scientists use plots to notice patterns that are difficult to see in raw numbers.

1.4.6 Reading and Interpreting Code

In a script, you read code from top to bottom, and you need to understand what each line does. Read code one line at a time. After each line, ask: "What changed?" Each line builds on the previous one. New lines of code can change objects created earlier, and functions may be called that transform data. The result from one line is often used again in the next line. This is why it's important to read code carefully and understand the structure of objects and functions.

Here is a short example:

```r
result <- mean(c(10, 12, 14, 16))
result
```

```
[1] 13
```

Read the code from the inside out and ask yourself:

1. What object (such as a vector, list, or data frame) is being created or modified? (Here: a vector of four numbers.)
2. What function (e.g. mean, for loop) is being used? (Here: `mean()`.)
3. What data are being selected or transformed?
4. What result should appear?

Let's break it down from the inside out:

- `c(10, 12, 14, 16)` creates a vector containing four numbers.
- `mean(...)` calculates the average of those numbers.
- `result <-` stores the average in an object called `result`.
- `result` displays the value stored in the object, which is the average of the numbers: 13.

> **i Focus on the ideas**
>
> You are not expected to memorize everything in this chapter.
> The goal is to:
>
> - become comfortable seeing code,
> - understand the basic ideas,
> - and recognize common patterns.
>
> These concepts will appear many times later in simpler and more practical examples.

1.5 Exercises

Try the following exercises using the Earth-data concepts we just covered:

1. Create a vector called `temperatures` that contains three temperature values (in °C).
2. Calculate the mean temperature.
3. Create a data frame with three columns: `station`, `temp`, and `rain` (similar to the data-frame example earlier in this chapter).
4. Extract the first row of this data frame.
5. Write a `for` loop that prints the numbers 1 to 3.

1.6 Key Takeaways

- R works through objects and functions.
- Scripts are better than console-only work because they can be saved and reused.
- Vectors, data frames, lists, and arrays are data structures.
- Reading code carefully is just as important as writing code.
- These ideas will be used again when working with spatial and time-based data.

So far, we have focused on the basic tools of R: objects, vectors, functions, and data frames. These tools allow us to write code and understand how R works. But real analysis in Earth science does not begin with clean, simple examples; it begins with real data, often complex, incomplete, and shaped by natural processes.

The goal of this chapter was not to memorize commands, but to understand how R organizes and processes data. These ideas become meaningful when we apply them to real Earth datasets, where patterns represent physical processes such as climate variability, groundwater movement, or tectonic change.

In the next chapter, we take that step. You will begin working with actual Earth and environmental data from the United States Geological Survey (USGS): reading it into R, inspecting its structure, and preparing it for analysis. This step is important because good analysis begins with understanding your data. The quality of any result, whether a map, a time series, or a model, depends on how well the data are understood and organized before analysis begins.

R is not the goal. It is one of many tools used in Earth data analytics. The goal is to develop the ability to think critically about data, recognize patterns, and connect those patterns to real-world Earth, environmental, and sustainability processes. You are not expected to master computer science concepts. Instead, you are building practical skills that allow you to work with data, ask meaningful questions, and interpret results in an Earth science context.

1.6.1 References (Data and Further Reading)

1. R Project for Statistical Computing. https://www.r-project.org

2. RStudio / Posit. https://posit.co

3. Wickham, H., Çetinkaya-Rundel, M., & Grolemund, G. *R for Data Science.* https://r4ds.hadley.nz

4. maps package documentation. https://cran.r-project.org/package=maps

Chapter 2

Data Handling and Preparation

Working with and Preparing USGS Groundwater Data in R

2.1 Introduction

In Chapter 1, we learned how to create and work with objects (named places to store things), vectors (a group of values stored together), functions (commands that do a job), and data frames (tables with rows and columns). In this chapter, we begin using those same tools with real-world datasets. Before data can be analyzed, visualized, or modeled, it must first be understood and prepared carefully.

Data preparation is one of the most important parts of data analytics. Small problems in the data can lead to misleading results later. Cleaning and checking data carefully helps make analyses reproducible, reliable, and scientifically meaningful.

In geoscience, real data are rarely perfect. You might see:

- missing values (gaps),
- repeated rows,
- confusing column names

that need careful preparation before analysis. Before we can map, model, or interpret a dataset, we need to make sure the data are organized in a way that we can trust and work with.

We use a USGS (United States Geological Survey) groundwater example here because it is a real environmental dataset that naturally contains missing values, making it a useful teaching example for beginners. The principles we cover here apply to any time series dataset, not just groundwater records. The same retrieval workflow scales to much larger studies. Buchbinder et al. (2021) used R to download and clean about 238,000 groundwater records from 56 USGS monitoring stations across Long Island, NY, then examined how stay-at-home orders during the COVID-19 pandemic changed water use.

> **i** What you will be able to do
>
> By the end of this chapter, you will be able to:
>
> - download real groundwater data from the USGS National Water Information System into R
> - inspect a dataset's structure and check its quality with `head()`, `names()`, `summary()`, and `str()`
> - keep and rename the columns you need to build a tidy table
> - find missing values and missing dates, then align the data by building a complete date sequence and merging on date
> - estimate a short gap by linear interpolation, and judge when filling a gap is reliable and when it is not
> - merge two real datasets (groundwater and precipitation) by date and rescale them onto a common 0-to-1 scale to compare patterns

2.1.1 What Are We Doing in This Chapter?

Before we start coding, it helps to understand the workflow. The chapter focuses on preparing data so you can trust it and use it: cleaning it, checking for issues, and organizing its structure. Careful preparation ensures our later analysis is based on reliable and well-understood data. In simple terms, data handling follows this process:

1. Load the data
2. Look at the structure
3. Clean the data
4. Check for problems (missing values, duplicates)
5. Fix or understand those problems
6. Prepare the data for analysis

These steps follow a standard analytical workflow used in research and industry:

load → inspect → clean → validate

This workflow helps ensure that later analysis is based on data that are organized, consistent, and scientifically interpretable.

At this stage, we are not trying to interpret the system yet; we are preparing the data so later analysis becomes reliable and meaningful.

It is like cleaning laboratory equipment before an experiment. If the equipment is dirty, the results can be wrong; in our case, even good analysis can produce misleading results.

How to Use This Chapter

- If you are new → follow Part A only
- If you feel comfortable → continue to Part B
- If you want a challenge → try Part C

Beginners can complete Part A and still build a solid foundation.

2.2 Part A: Beginner (Must Follow)

2.2.1 Reading and Inspecting the Data

We are using a function (a ready-made command that does a job for us) to download data from the USGS website. For now, focus on what the data look like after loading. The key point is that we are working with real data that may have issues, and we need to understand those issues before we can analyze the data.

> **i Focus on the workflow**
>
> We are downloading real groundwater data, storing it in an object (a named container that holds the data so we can use it again), and then looking at the results. You will gradually become more comfortable reading longer code blocks.
> Note: This section uses the `dataRetrieval` package (a package is an add-on toolbox of extra commands). If you skipped Part B of Chapter 1, now is a good time to read the section that explains how `library()` switches a package on and what to do if a package is not yet installed. The `# install.packages("dataRetrieval")` line at the top of the code block is how you install it the first time.
> You must be connected to the internet for this block to work, because it downloads the data live from the USGS website. If you see a red error that mentions the package, run `install.packages("dataRetrieval")` once first, then run the block again.

```r
# Uncomment the line below only if you need to install the package
# install.packages("dataRetrieval")
library(dataRetrieval)

# One groundwater station for a beginner-friendly example.
# This 15-digit number is the USGS ID for one specific monitoring well;
# you can look up stations for your own area on the USGS NWIS website.
# In quotes, so R treats it as text (a "character" label),
# not a number to do math on
station_id <- "404416073491801"
# USGS parameter code for groundwater level (ft above NGVD 1929)
parameter <- "62610"

start_date <- "2000-02-01"
end_date   <- "2025-02-10"

# The name = value pairs below are "arguments":
# the settings we hand to the function.
```

```r
gw_raw <- readNWISdv(
  siteNumbers = station_id,
  parameterCd = parameter,
  startDate   = start_date,
  endDate     = end_date
)

# head() shows just the first few rows so we can quickly view the data.
head(gw_raw)
```

```
  agency_cd         site_no       Date X_62610_00003 X_62610_00003_cd
1      USGS 404416073491801 2006-09-16          7.59                A
2      USGS 404416073491801 2006-09-17          7.62                A
3      USGS 404416073491801 2006-09-18          7.68                A
4      USGS 404416073491801 2006-09-19          7.73                A
5      USGS 404416073491801 2006-09-20          7.76                A
6      USGS 404416073491801 2006-09-21          7.75                A
```

The dataset is a table, and each column is one labelled strip of values running down the table (like a column in a spreadsheet). Some column names are hard to read (for example: X_62610_00003), and the dataset also includes columns ending in `_cd`, which are quality codes you can ignore for now. We will rename the important columns so they are easier to read.

> **Best practice for reproducibility**
>
> This chapter downloads **live** data directly from the USGS National Water Information System using `readNWISdv()` with `endDate = "2025-02-10"`. After downloading the data the first time, save the raw version locally so you can reproduce the exact results later:
>
> ```r
> write.csv(gw_raw, "data/usgs_gw_raw.csv", row.names = FALSE)
> ```

The `names()` function lists every column heading in the table, so we can see what we have to work with.

```r
names(gw_raw)
```

```
[1] "agency_cd"        "site_no"          "Date"             "X_62610_00003"
[5] "X_62610_00003_cd"
```

The date is stored in a column named `Date`. The site number is stored in a column named `site_no`. These are the key variables we will work with.

Now that we have loaded and inspected the groundwater dataset, the next step is to simplify and organize it. Real-world datasets often contain many columns, technical variable names, and extra metadata that are not immediately useful for analysis. Cleaning the data helps us focus on the variables we actually need.

2.2.2 Cleaning the Data

A common first step is to keep only the columns we need. We can then rename them using shorter and more meaningful names. When we want to keep more than one column, we can use `%in%`, which means "is one of these values."

> **The %in% operator**
>
> `%in%` checks whether each value on its left side appears in the list on its right side. For each value it answers with `TRUE` (yes, it's in the list) or `FALSE` (no, it isn't).

```
"Nassau" %in% c("Kings", "Queens", "Nassau", "Suffolk")  # TRUE
"Albany" %in% c("Kings", "Queens", "Nassau", "Suffolk")  # FALSE
```

The key variables are the site number, the date, and the groundwater level.
The groundwater-level column may have a name like `X_62610_00003`, which is hard to read. This column name is specific to the station and parameter code used in this example. If you use a different station or parameter, the column name will be different (the number in it changes with the parameter code), but the cleaning steps below will still apply.

These values are heights in feet above NGVD 1929, a fixed national reference level first set from average sea level (it was originally called the "Sea Level Datum of 1929"). Roughly, 0 is about average sea level, so a reading near 7.6 means the water table sits about 7.6 feet above that reference.

> **A small selection pattern**
>
> - `c()` puts several values together into one group (the "c" stands for *combine*).
>
> Here, it is used to create a group of column names that we want to keep.

Think of a data frame like a spreadsheet. Sometimes real datasets contain dozens of columns, but we may only need a few of them. Here, we are telling R:
"Keep only these three columns from the table."
Inside the square brackets, the part before the comma chooses rows and the part after the comma chooses columns. Leaving the row part blank means "keep all rows."

```r
# Keep only the columns we need
gw_raw_small <-
  gw_raw[, names(gw_raw) %in% c("site_no", "Date", "X_62610_00003")]
```

The above code may look complicated at first, but the main idea is simple:

- `names(gw_raw)` gets all the column names in the dataset.
- `c("site_no", "Date", "X_62610_00003")` creates a group containing the three column names we want to keep.
- `%in%` checks which column names belong to that group.
- `[ ]` keeps only the matching columns.

As a result, R creates a smaller table containing only the columns we need.

You can think of it like highlighting only the spreadsheet columns you want to keep and hiding the rest.

```r
# Build a simple working data frame (a table of our own).
# The $ sign picks out one column from a table: table$column.
# as.Date() tells R "treat this as a calendar date";
# as.numeric() "treat this as a number."
gw <- data.frame(
  site_no = gw_raw_small$site_no,
  date    = as.Date(gw_raw_small$Date),
  gwl     = as.numeric(gw_raw_small$X_62610_00003)
)

head(gw)
```

```
          site_no       date  gwl
1 404416073491801 2006-09-16 7.59
2 404416073491801 2006-09-17 7.62
3 404416073491801 2006-09-18 7.68
4 404416073491801 2006-09-19 7.73
5 404416073491801 2006-09-20 7.76
6 404416073491801 2006-09-21 7.75
```

```r
summary(gw)
```

```
   site_no               date                 gwl
 Length:3385        Min.   :2006-09-16   Min.   : 3.24
 Class :character   1st Qu.:2010-02-21   1st Qu.: 8.90
```

```
 Mode  :character   Median :2017-03-19   Median :11.18
                    Mean   :2015-05-29   Mean   :10.92
                    3rd Qu.:2019-10-28   3rd Qu.:13.25
                    Max.   :2022-02-22   Max.   :16.70
```

We also rename the columns while creating the new data frame to make the dataset easier to work with. Clear column names make it easier to understand what each variable represents. Here, `gwl` stands for groundwater level. The `summary()` function gives us a quick overview of the data, including the range of dates, the range of groundwater values, and the presence of missing values. In the printed output, the `Min.` and `Max.` rows are the smallest and largest values, `Median` is the middle value, and `1st Qu.` and `3rd Qu.` mark the quarter points (a quarter and three quarters of the way through the sorted values). If a column had any missing values, an `NA's` row would also appear. In R, a missing value is shown as `NA`, which simply means "no value recorded here."

> ⚠ Avoid using reserved names
>
> Avoid naming objects with common function names such as `data`, `list`, or `mean`. These names already exist in R and can cause confusing errors.

2.2.3 Exploring the Data

Before detecting any missing data, it helps to see the raw time series. This may look like a lot of code, but focus on the idea: we are plotting date along the bottom (the x-axis) and groundwater level up the side (the y-axis).

```
# The short settings control the look: pch = point shape, col = color,
# lwd = line width, xlab/ylab = axis labels, main = title. grid() adds
# gridlines.
plot(
  gw$date,
  gw$gwl,
  pch = ".",
  col = "black",
  lwd = 1,
  xlab = "Date",
  ylab = "Groundwater level (ft above NGVD 1929)",
  main = "USGS Groundwater Record"
)
grid()
```

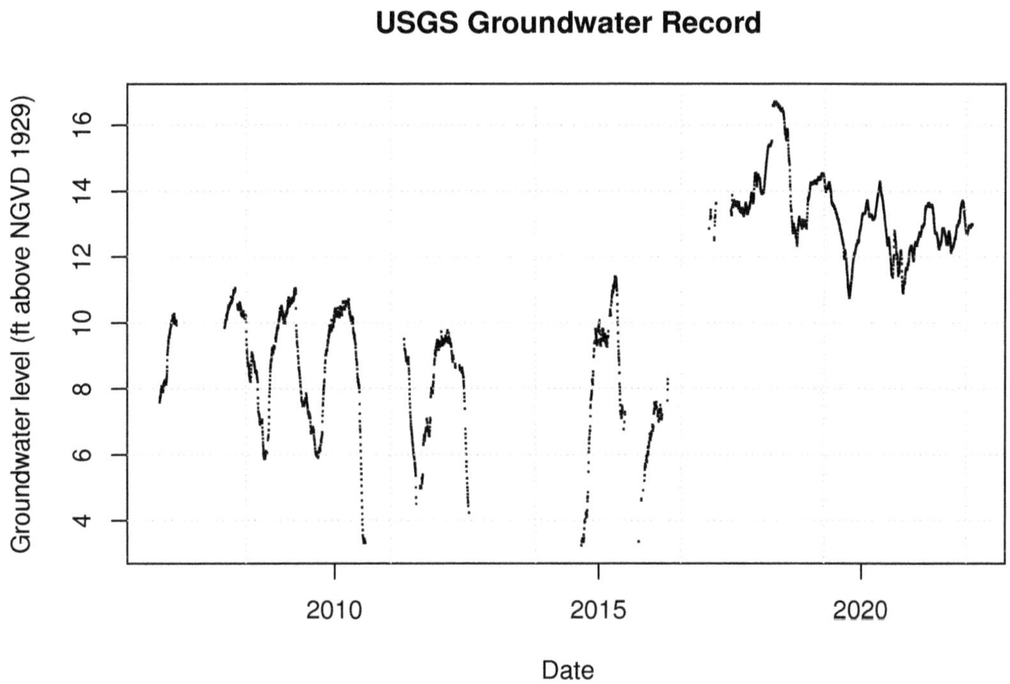

Figure 2.1: Raw USGS groundwater time series showing gaps due to missing observations.

This simple plot, shown in Figure 2.1, helps us ask three questions:

- Does the line move smoothly from left to right? (correct date order)
- Do you see breaks? (missing data)
- Does the line look smooth or noisy?

Even a basic plot can reveal problems before we begin any analysis. For example, if the dates are not in order, we will need to sort them before doing any time series analysis. Large gaps mean we must decide how to handle those missing values, and an unusual pattern may signal outliers or measurement errors that need investigation.

At this point, you can already load, inspect, and visualize real data. That is enough to begin basic analysis. Next, we will look at common problems in real data, such as missing values and unmatched dates.

2.3 Part B: Intermediate (Recommended)

Now we move from seeing data to fixing problems in data. We will detect and handle missing values, check for duplicates, and prepare the data for analysis. This step helps ensure that later analysis is reliable and interpretable.

2.3.1 Detecting Data Issues

Two common problems are missing values and duplicated rows. We first check if any values are missing.

```
# Read nested commands from the inside out: is.na(gw) runs first (marks
# each cell
# TRUE if missing), then colSums() adds up the TRUEs in each column.
colSums(is.na(gw))
```

```
site_no    date     gwl
      0       0       0
```

```
# Duplicate dates
sum(duplicated(gw$date))
```

```
[1] 0
```

`is.na()` checks for missing values, and `colSums()` counts how many missing values exist in each column. The result shows one number per column; here every number is `0`, so no values are missing inside the rows we already have. The duplicate check on the next line prints a single number, and the `0` confirms no date appears twice. However, `is.na()` can only detect missing values in rows that already exist. If an entire date is missing from the dataset, there is no row to contain an `NA` value.

The `duplicated()` function checks whether the same date appears more than once. There are no duplicated dates in this example, but if there were, we would need to investigate whether they represent true duplicates or whether they contain different values that need to be resolved. In some cases, we may keep one of the duplicates (e.g. the maximum or the average), but not both.

Missing values are not automatically bad. In groundwater records, they may reflect instrument gaps, maintenance, or days when no measurement was available. However, we need to be aware of them and decide how to handle them before performing any analysis.

There are two different kinds of gaps:

- we have the date, but the value is missing
- the entire date is missing from the table

Both types can create gaps in the time series, but they require different handling.

To identify missing dates, we create a complete sequence of expected dates and compare it with the dates that actually exist in the dataset.

Before writing code, think about the problem. Imagine a calendar on the wall. If the groundwater station worked perfectly, every day on the calendar would have a measurement. Real monitoring stations sometimes stop recording temporarily, creating gaps.

```
# Find the earliest date
start_date <- min(gw$date)

# Find the latest date
end_date <- max(gw$date)

# Create a complete sequence of dates
date_seq <- seq(start_date, end_date, by = "days")

# Check for missing dates: which expected dates are NOT in our records?
# The as.Date() wrapper turns the result back into proper calendar dates.
missing_dates <- as.Date(setdiff(date_seq, gw$date))
```

The `setdiff()` function finds values that appear in the first object but not in the second. In other words, `setdiff(list_A, list_B)` returns everything in list A that does not appear in list B.

Here, it finds dates that appear in the complete daily sequence (`date_seq`) but do not appear in the USGS record.

In other words, `setdiff()` means:

"Show me what exists in the full calendar, but does NOT exist in my dataset."

```
length(missing_dates)
```

```
[1] 2254
```

The `length()` function counts how many missing dates were found. If the result is 0, then no dates are missing from the time series.

In other words:

- `date_seq` = all expected dates
- `gw$date` = dates actually recorded
- `setdiff()` finds the missing ones

If there are missing dates, we can decide whether to add rows with `NA` values for those dates or to leave them out, depending on the context of the analysis.

Imagine two spreadsheets:

- one spreadsheet contains all dates
- the second spreadsheet contains only dates with measurements

We now combine them by matching rows that share the same date. If a date exists in the calendar but not in the groundwater measurements, the groundwater value becomes `NA` (missing).

```
# Create a complete data frame with all dates
gw_complete <- data.frame(date = date_seq)
```

The `merge()` function combines the two tables using the date column. You can think of it like matching rows between two spreadsheets using the same date.

```
# Here "x" is the first table listed (gw_complete, our full calendar).
merged_gw <- merge(gw_complete, gw, by = "date", all.x = TRUE)
head(merged_gw)
```

```
        date         site_no  gwl
1 2006-09-16 404416073491801 7.59
2 2006-09-17 404416073491801 7.62
3 2006-09-18 404416073491801 7.68
4 2006-09-19 404416073491801 7.73
5 2006-09-20 404416073491801 7.76
6 2006-09-21 404416073491801 7.75
```

```
# Confirm that the number of rows matches
# the length of the date sequence
nrow(merged_gw) == length(date_seq)
```

```
[1] TRUE
```

```
# Confirm that no dates are missing in the merged data frame:
# subtracting the dates in merged_gw
# from the original date sequence should yield an empty set
setdiff(date_seq, merged_gw$date)
```

```
numeric(0)
```

In the output, `TRUE` confirms the row counts match, and the empty result (`character(0)` or `integer(0)`) means no dates are missing.

Merging takes place by combining two tables using a shared column (here: date). `all.x = TRUE` tells R:

"Keep the full calendar of dates, even when groundwater measurements are missing."

If there is no matching date in the original data, the groundwater level will be filled with `NA`, indicating a missing value for that date.

It is useful to plot the missing values to see where they occur in the time series. This can help us understand whether the gaps are random or if they cluster in certain periods, which may have implications for how we handle them.

```
# Let's plot the original data
# with missing values highlighted
plot(
  merged_gw$date,
  merged_gw$gwl,
  type='l',
  col = "black",
  lwd = 1,
  xlab = "Date",
  ylab = "Groundwater level\n (ft above NGVD 1929)",
  main = "Groundwater Record with Missing Values"
)
abline(v = merged_gw$date[is.na(merged_gw$gwl)],
       col = "pink", lwd = 0.5)

# Also overlay the raw data to
# visualize the missing points more clearly
points(gw$date, gw$gwl, pch = ".", col = "black", lwd = 1)
```

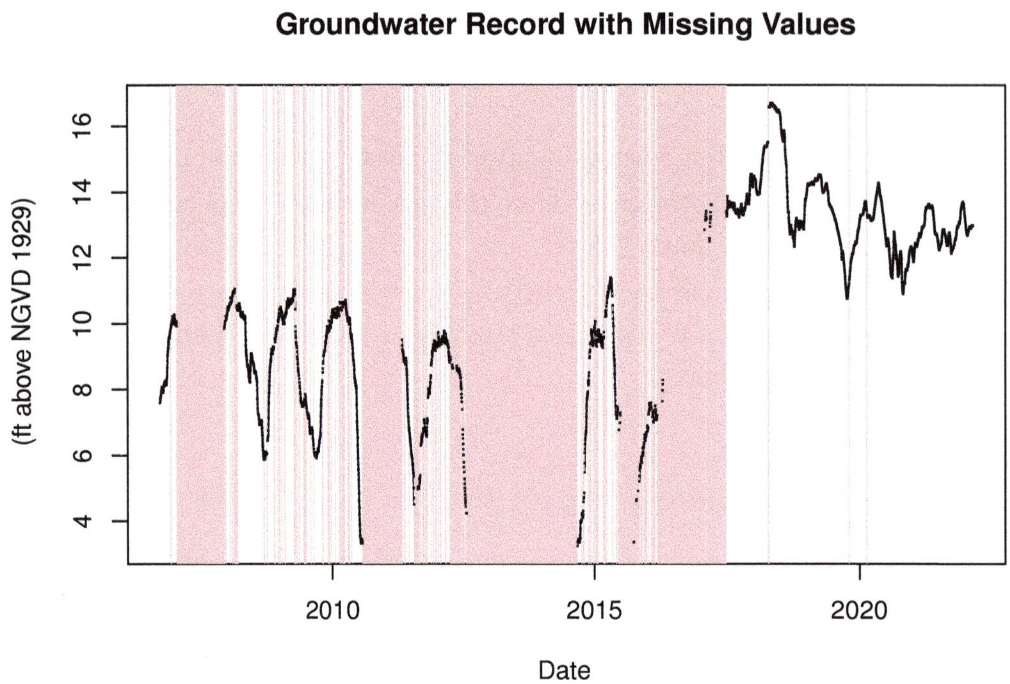

Figure 2.2: Groundwater record with missing values highlighted in pink.

Missing observations occur at several points throughout the record (Figure 2.2). Identifying these gaps is important because missing values can affect later analyses such as trend detection, interpolation, and modeling. Visualizing missing observations also helps communicate data quality in scientific reports, publications, and conference presentations. It allows us to see the structure of the gaps and determine whether missing values occur as isolated observations or as longer periods of missing data.

2.3.2 Handling Missing Data

Now we count missing groundwater values.

```
sum(is.na(gw$gwl))
```

```
[1] 0
```

The result is `0`, which means there are no missing groundwater values (`NA`) in the original dataset. However, this does not mean that the record is complete. Earlier, we found that some dates are entirely missing from the time series. Those missing dates become `NA` values only after we create a complete sequence of dates and merge the groundwater observations onto it.

Time series should always be sorted by date because many functions assume the data are in chronological order. If the data are not sorted, the results can be misleading.

We now introduce a simple interpolation example. The goal is only to see how a missing value can be estimated from nearby observed values.

```
# install.packages("zoo") # Uncomment if needed
library(zoo)

# Make sure the merged data are sorted by date
merged_gw <- merged_gw[order(merged_gw$date), ]
```

2.3.3 Simple Interpolation Example

In this example, we will show one small gap so you can see how linear interpolation works. We will pick one missing value and look at it closely. This allows us to understand how interpolation works without worrying about the size of the gap. Interpolation is a way to guess a missing value using nearby known values. Imagine drawing a straight pencil line between two dots on paper. If one point in the middle is missing, interpolation estimates where that missing point probably belongs along the line. The goal here is only to understand the idea, not to decide whether a gap is safe or risky.

```
# An "index" is just a row's position number (row 1, row 2, ...).
# which() reports the position numbers of the missing values.
na_idx <- which(is.na(merged_gw$gwl))

# na_idx[1] picks the first of those positions (the [1] means "item number 1").
i <- na_idx[1]

# Create a small window around that point
window_idx <- (i-5):(i+5)
window_idx <- window_idx[window_idx > 0 &
                           window_idx <= nrow(merged_gw)]

simple_df <- merged_gw[window_idx, ]

# Apply linear interpolation
simple_df$gwl_interp <- zoo::na.approx(simple_df$gwl)

# Plot
plot(
  simple_df$date,
  simple_df$gwl,
  pch = 20,
```

```r
  col = "lightgreen",
  cex = 2,
  xlab = "Date",
  ylab = "Groundwater",
  main = "Simple Interpolation Example"
)

points(simple_df$date, simple_df$gwl_interp,
       col = "red", pch = 4, cex = 1.5)

legend(
  "topleft",
  legend = c("Observed", "Interpolated"),
  col = c("lightgreen", "red"),
  pch = c(20, 4),
  bty = "n"
)
```

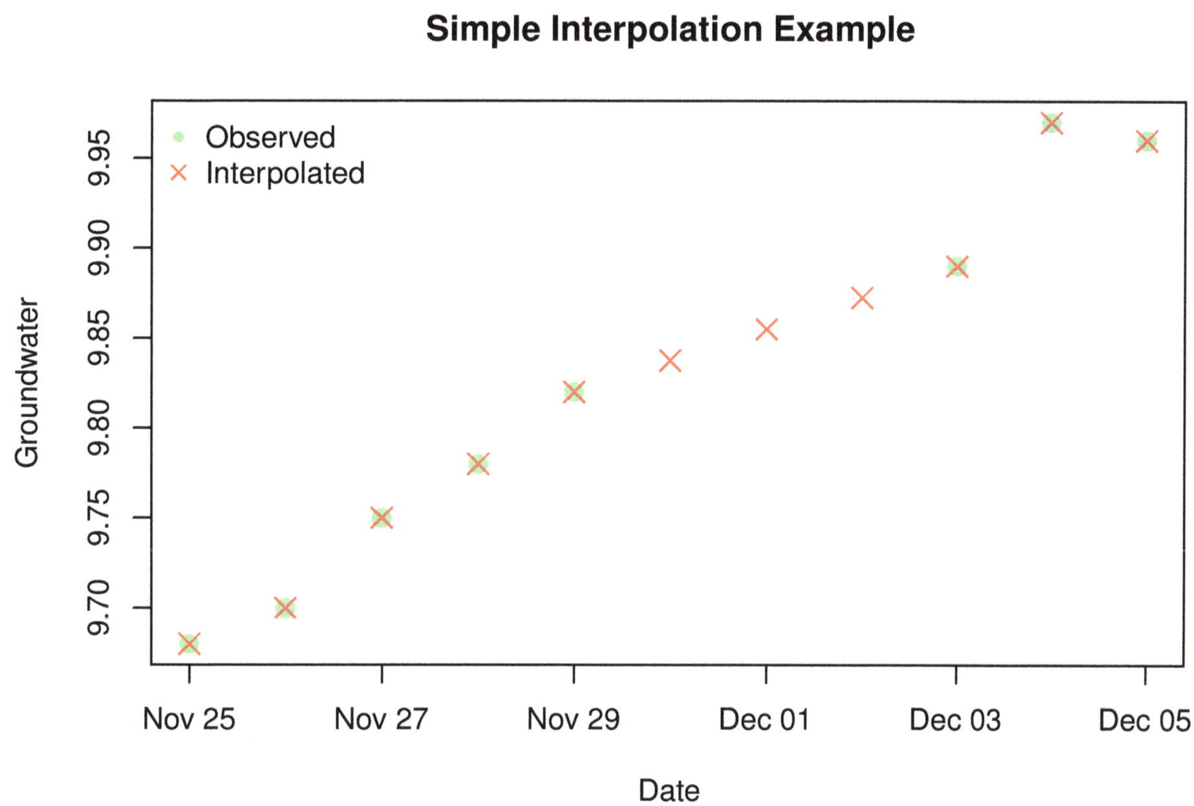

Figure 2.3: A small-gap example: linear interpolation estimates one missing groundwater value from the nearby observed points.

The light green points are the real measured values; the red X marks are the values the computer estimated to fill the gap. Where a red X sits between green dots, that point was guessed, not measured. In Part B, we are only learning how interpolation works. In Part C, we will look at when it is safe to trust it.

2.4 Part C: Advanced (Optional Challenge)

Now that we understand how interpolation works, we can ask a more important question: when should we trust it? To answer this, we will distinguish between short gaps and long gaps, and we will build a way to detect them automatically. Short gaps are often reasonable to estimate. Long gaps are riskier because they may hide real changes.

This section introduces a more flexible and automated workflow for detecting and handling missing data. We will write code that can automatically detect gaps, choose which ones to interpolate, and apply interpolation only where it makes sense. This is more advanced but helps create a more reliable data handling workflow.

2.4.1 Looking Deeper at Data Structure

The function `str()` shows the structure of a dataset: for each column it lists the column name, the kind of data it holds (text, numbers, dates), and a few example values. This is often the fastest way to understand what each column contains and how R is storing the data.

The function `attributes()` shows metadata attached to the dataset, such as source information, units, or additional descriptions.

In this example, the groundwater values are stored in a column with a technical name that includes the parameter code. The date is stored in a column named `Date`, and the site number is stored in a column named `site_no`.

These are the key variables we will work with as we clean and analyze the data.

```
# str(gw_raw) prints a very long, messy block because the dataset carries extra
# hidden metadata (units, source notes). We copy it and remove that metadata so
# the structure prints cleanly. Running str(gw_raw) yourself still works;
# it just
# shows the same columns plus a lot of extra attribute text.
gw_removedattributes <- gw_raw
attributes(gw_removedattributes) <- NULL
# vec.len = 2 prints only 2 example values per column to save space
str(gw_removedattributes, vec.len = 2)
```

```
List of 5
 $ : chr [1:3385] "USGS" "USGS" ...
 $ : chr [1:3385] "404416073491801" "404416073491801" ...
 $ : Date[1:3385], format: "2006-09-16" "2006-09-17" ...
 $ : num [1:3385] 7.59 7.62 7.68 7.73 7.76 ...
 $ : chr [1:3385] "A" "A" ...
```

We first inspect the column names to see what we are working with.

The groundwater-level column may have a name like X_62610_00003, whose meaning is not obvious. Instead of typing the column name manually, we can find it automatically. We use the function grep() to find the column name that contains the parameter code. This way, we can write code that works for any parameter without typing the exact column name.

```
# paste0() joins text together to build "X_62610";
# grep() then searches the column
# names for that text and returns the matching name; [1] takes the first match.
value_col <- grep(paste0("X_", parameter),
                  names(gw_raw), value = TRUE)[1]
# Rebuild the gw table from Part A, this time using the column name found
# above.
# gw_raw[[value_col]] pulls out one column by the name stored in value_col.
gw <- data.frame(
  site_no = gw_raw$site_no,
  date    = as.Date(gw_raw$Date),
  gwl     = as.numeric(gw_raw[[value_col]])
)

head(gw)
```

```
          site_no       date  gwl
1 404416073491801 2006-09-16 7.59
2 404416073491801 2006-09-17 7.62
3 404416073491801 2006-09-18 7.68
4 404416073491801 2006-09-19 7.73
5 404416073491801 2006-09-20 7.76
6 404416073491801 2006-09-21 7.75
```

```
summary(gw)
```

```
   site_no               date                 gwl
 Length:3385        Min.   :2006-09-16   Min.   : 3.24
 Class :character   1st Qu.:2010-02-21   1st Qu.: 8.90
 Mode  :character   Median :2017-03-19   Median :11.18
                    Mean   :2015-05-29   Mean   :10.92
                    3rd Qu.:2019-10-28   3rd Qu.:13.25
                    Max.   :2022-02-22   Max.   :16.70
```

Now that we have seen a basic interpolation example, we can look at a more advanced question: how do we identify short and long gaps automatically?

2.4.2 Automatic Gap Detection

Not all missing data should be treated the same way. The decision depends on the structure of the data and the length of the gap.

Example 1: Short Gap Interpolation

Because the missing values are surrounded by nearby observations, a short gap is where interpolation is most likely to provide a reasonable estimate. We will use linear interpolation, which connects the known points with a straight line. This method assumes that the change between observations is smooth and gradual, which is often reasonable for groundwater levels over short periods.

If the missing gaps are short, interpolation may provide a reasonable estimate. If the gaps are long, interpolation may be misleading. To understand this difference, we examine two cases from the same dataset: a short gap and a long gap. In the previous figure, the pink vertical lines showed where data are missing. Here, we will look closely at those gaps and examine how interpolation behaves in each case.

First, we will focus on a short gap, where interpolation can provide a reasonable estimate between nearby observations. Then, we will examine a long gap, where interpolation creates a smooth line that may look convincing but does not represent real measurements.

Even short-gap interpolation relies on assumptions that may not always be true in real Earth systems.

Now that we understand what a gap is, we can write code to find short and long gaps automatically. We use the `rle()` function to identify consecutive missing values, then summarize the start, end, and length of each gap in a data frame. This helps us distinguish between short gaps that may be suitable for interpolation and longer gaps that require more caution.

The next code may look complex. Focus on the idea: we are finding where gaps occur and how long they are.

```
# Identify consecutive missing values
na_rle <- rle(is.na(merged_gw$gwl))
```

`rle()` stands for "run length encoding." It scans through the values and reports each unbroken streak of the same value (for example, "5 missing values in a row, then 12 present, then 3 missing"). Here it lets us find runs of missing values (`NA`) and measure how long each gap lasts.

After identifying the missing sections, we now calculate:

- where each gap starts,
- where it ends,
- and how many missing days it contains.

```
# cumsum() = running total. Adding up the streak lengths in order
# tells us the row number where each streak ENDS; subtracting the
# streak's own length (and adding 1) gives the row where it STARTS.
# This just converts streak sizes into row positions.
na_ends <- cumsum(na_rle$lengths)
na_starts <- na_ends - na_rle$lengths + 1

# Create a data frame to summarize the gap information
# of start, end, and length of each gap
gap_info <- data.frame(
  start = na_starts[na_rle$values],
  end   = na_ends[na_rle$values],
  n     = na_rle$lengths[na_rle$values]
)

# Choose the shortest gap as a simple interpolation example
short_gap <- gap_info[which.min(gap_info$n), ]

# Distribution of gap lengths
table(gap_info$n)
```

```
  1   2   3   4   5   6   7   8   9  10  13  16  20  26  37  92 102 270 284 329
 40  44  16   5   3   2   1   1   2   1   1   1   1   2   1   1   1   1   1   1
773
  1
```

```
# Peek at the first few gaps to see what the code produced
head(gap_info)
```

```
  start end   n
1    76  78   3
2   119 447 329
3   457 459   3
4   490 491   2
5   504 506   3
6   516 518   3
```

> Reading this output
>
> **gap_info** has one row per gap, with **start** and **end** giving the row positions where the gap begins and finishes, and **n** giving how many days are missing. The **table(gap_info$n)** line above counts how many gaps have each length: the top row lists the gap lengths, and the bottom row lists how many gaps have that length. You do not need to follow every line of the code; the goal is just to know where the gaps are and how long they last.

Most gaps are short, while only a few are very long. Short gaps are often more suitable for interpolation than long gaps.

```
# We will keep a small window around that gap
# to show the interpolation in context.
short_idx <- seq(
  max(1, short_gap$start - 10),
  min(nrow(merged_gw), short_gap$end + 10)
)

short_df <- merged_gw[short_idx, ]

# Linear interpolation in this short window
short_df$gwl_interp <- zoo::na.approx(
  short_df$gwl, # Interpolate missing groundwater values
  x = as.numeric(short_df$date),
  na.rm = FALSE
)
par(mar=c(5,5,5,5))
plot(
  short_df$date,
  short_df$gwl,
  type = "p",
  pch = 20,
  cex = 3.5,
  cex.lab = 0.9,  # Adjust axis label size
  cex.main = 0.9, # Adjust main title size
  cex.axis = 0.8, # Adjust axis numbers size
  col = "lightgreen",
  xlab = "Date",
  ylab = "Groundwater level\n (ft above NGVD 1929)",
  main = "Short Gap: Linear Interpolation Example"
)
```

```
points(short_df$date, short_df$gwl_interp, col = "red", lwd = 2, pch="X")

legend(
  "topleft",
  legend = c("Observed values", "Linear interpolation"),
  pch = c(20, 4),          # 20 = filled circle, 4  "X"
  col = c("lightgreen", "red"),
  cex = 0.7, # legend text size
  pt.cex = c(3.5, 1.5),    # match point sizes (optional but nice)
  bty = "n"
)
```

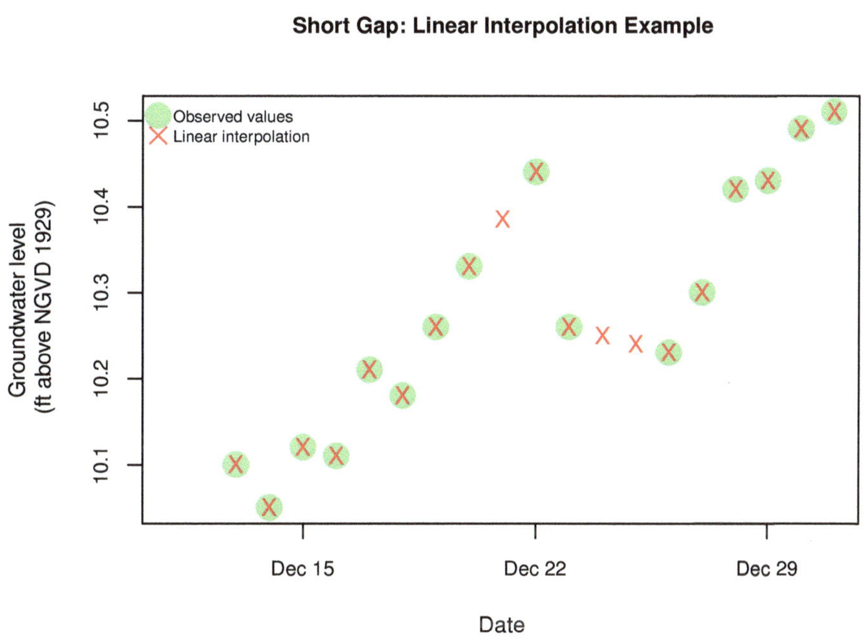

Figure 2.4: Short gap interpolation. Estimated values (red X) follow nearby observations closely.

Interpolation works better in short gaps (Figure 2.4). In this example, the missing values occur over a short time period, and the surrounding observations are close in both time and magnitude. The interpolated values (red X marks) follow a smooth transition between the observed points, which suggests that the estimate is reasonable. The linear interpolation follows the general trend of the groundwater level across the gap without introducing unrealistic jumps or distortions. In this case, using linear interpolation to fill the missing values is a simple and practical way to close small gaps without significantly distorting the overall pattern.

Example 2: Long Gap Interpolation

Now we move to a more challenging situation: a long gap in the data. Unlike the previous example, the missing values span a much larger period, meaning there is less information to guide what happens between the observed points. In this case, applying interpolation becomes more questionable. Even though a mathematical line can still be drawn, it may not reflect the true behavior of the groundwater system during the missing period.

```
# Choose the longest gap
long_gap <- gap_info[which.max(gap_info$n), ]

# Keep a small window around that gap
long_idx <- seq(
  max(1, long_gap$start - 10),
  min(nrow(merged_gw), long_gap$end + 10)
)

long_df <- merged_gw[long_idx, ]

# Interpolation is shown only to illustrate the problem
long_df$gwl_interp <- zoo::na.approx(
  long_df$gwl,
  x = as.numeric(long_df$date),
  na.rm = FALSE
)

par(mar=c(5,5,5,5))
plot(
  long_df$date,
  long_df$gwl,
  type = "p",
  pch = 20,
  cex = 3.5,
  cex.lab = 0.9,  # Adjust axis label size
  cex.main = 0.9, # Adjust main title size
  cex.axis = 0.8, # Adjust axis numbers size
  col = "lightgreen",
  xlab = "Date",
  ylab = "Groundwater level\n (ft above NGVD 1929)",
  main = "Long Gap: Interpolation Can Be Misleading"
)
```

```
points(long_df$date, long_df$gwl_interp, col = "red", lwd = 2, pch="X")

legend(
  "topright",
  legend = c("Observed values", "Linear interpolation"),
  pch = c(20, 4),         # 20 = filled circle, 4   "X"
  col = c("lightgreen", "red"),
  cex = 0.7, # legend text size
  pt.cex = c(3.5, 1.5), # match point sizes (optional but nice)
  bty = "n"
)
```

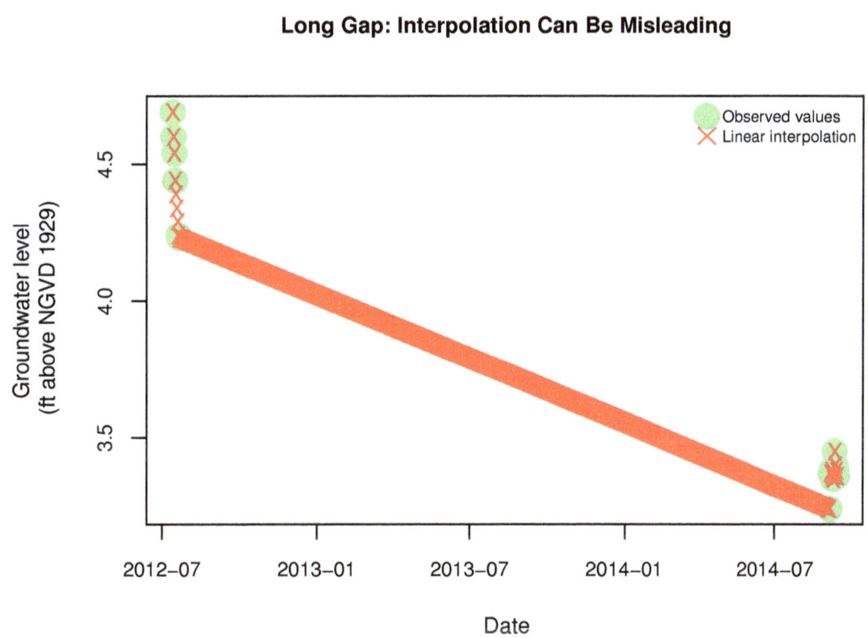

Figure 2.5: Long gap interpolation. The smooth line may not reflect real system behavior.

Interpolation over long gaps can be misleading (Figure 2.5). In this example, the gap is much longer, and there are no observations in between to guide the interpolation. The red X marks form a smooth connection between the two distant points, but this smooth transition is only a mathematical assumption.

In reality, the groundwater level may have increased, decreased, or fluctuated during this period. Because we do not have measurements, the interpolated values should not be interpreted as real data. This example shows why interpolation can be misleading for long gaps. Even though the result looks clean and continuous, it can hide important variability and give a false sense of accuracy.

For long gaps, it is often better to leave the values as missing or analyze the segments separately without trying to fill the gap.

> **Thinking Like an Analyst**
>
> Before filling missing values, ask:
>
> - Why are the data missing?
> - Is the gap short or long?
> - Could important environmental changes have occurred during the gap?
> - Will interpolation improve interpretation, or create misleading patterns?
>
> Good data analysis is not only about applying methods. It is also about understanding when those methods are scientifically appropriate.

> **Short vs. long gaps**
>
> Interpolation is most appropriate for short gaps where nearby observations suggest relatively stable behavior, and should be avoided or used with caution for long gaps where important changes may be hidden.
>
> - Short gaps → safe to estimate
> - Long gaps → risky (you might invent patterns that are not real)

> **Interpolation is an estimate**
>
> Interpolation estimates values within the gap. It does not recover the true missing observation.

The plot below shows the raw groundwater data and the interpolated series together.

```
# Choose the longest gap
par(mar=c(5,5,5,5))
plot(
  gw$date,
  gw$gwl,
  type = "p",
  pch = 20,
  cex = 1.2,
  cex.lab = 0.9,  # Adjust axis label size
  cex.main = 0.9, # Adjust main title size
  cex.axis = 0.8, # Adjust axis numbers size
```

```
  col = "lightgreen",
  xlab = "Date",
  ylab = "Groundwater level\n (ft above NGVD 1929)",
  main = "Long Gap: Interpolation Can Be Misleading"
)

points(long_df$date, long_df$gwl_interp, col = "red", lwd = 2, pch="X")

legend(
  "topleft",
  legend = c("Observed values", "Linear interpolation"),
  pch = c(20, 4),         # 20 = filled circle, 4    "X"
  col = c("lightgreen", "red"),
  cex = 0.7, # legend text size
  pt.cex = c(3.5, 1.5), # match point sizes (optional but nice)
  bty = "n"
)

grid()
```

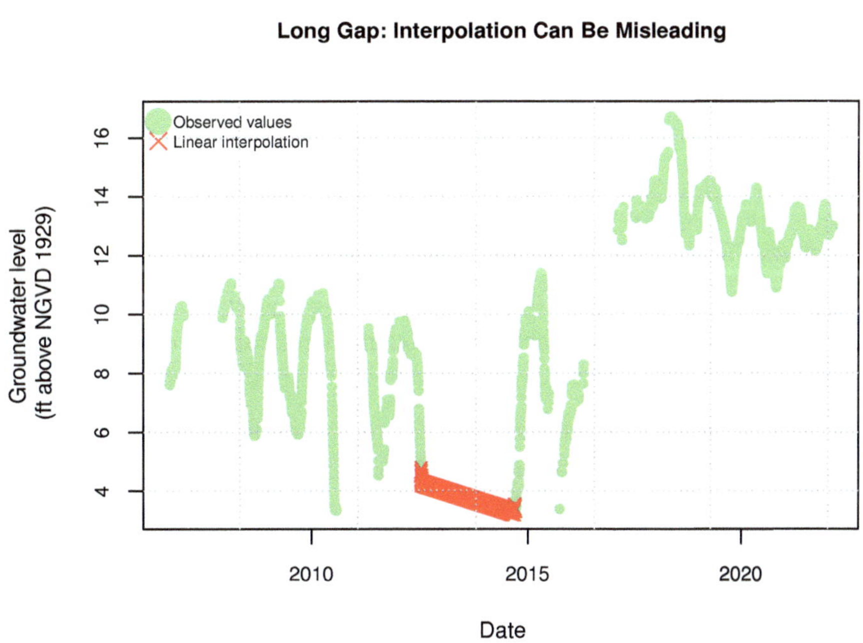

Figure 2.6: Long-gap interpolation example.

Interpretation: The light green points show the observed values (Figure 2.6). The red X marks show the interpolated values; they represent estimates, not actual observations.

These examples show that not all missing data should be treated the same way. The decision depends on the structure of the data and the length of the gap.

2.4.3 Common Mistakes to Avoid

⚠ Check for these issues

- Ignoring missing values and hoping they will not matter.
- Interpolating long gaps without checking whether it makes scientific sense.
- Treating interpolated values as measured values.
- Removing duplicated rows without checking whether they are true duplicates.
- Reordering time data incorrectly.

When is interpolation reasonable?

Short gaps + smooth behavior → interpolation is reasonable
Long gaps + unknown behavior → avoid or treat with caution

In real-world geoscience analysis, we often work with more than one dataset. For example, we may want to compare groundwater levels with precipitation, temperature, or another monitoring station.

To do this, we need to combine datasets in a consistent way. This requires careful handling of dates, alignment of observations, and sometimes rescaling variables so they can be compared meaningfully.

When working with precipitation, temperature, or other environmental datasets, we encounter many of the same challenges seen in groundwater records: missing values, duplicated observations, and inconsistent time coverage.

Before comparing datasets, we must inspect their structure, check for missing values, and ensure that the time series are properly aligned. These steps help make comparisons scientifically meaningful and reproducible.

ℹ Real data are rarely perfect

Unlike simulated data, real datasets may include missing values, measurement inconsistencies, or unit conventions that require attention.

2.4.4 Merging with Real Precipitation Data

In the previous example, we saw how datasets can be merged. Now we'll work with real environmental data by downloading daily precipitation from the Global Historical Climatology Network (GHCN) for JFK Airport. This allows us to combine two real geoscience datasets: groundwater levels and precipitation.

```
# install.packages("GHCNr") # Uncomment if needed
library(GHCNr)

# Like the first block, this downloads live data, so you must be
# online; if you see a package error, run install.packages("GHCNr")
# once first. USW00094789 is the GHCN station ID for JFK Airport,
# and "prcp" means precipitation.
# Download daily precipitation data
jfk_prcp <- daily(
  station_id = "USW00094789",
  start_date = "2000-01-01",
  end_date   = "2024-12-31",
  variables  = "prcp"
)

# Check the column names to understand
# the structure of the precipitation dataset
names(jfk_prcp)
```

```
[1] "date"       "station"    "prcp"       "prcp_flag"
```

The **names()** output lists the columns; the ones we need are the date column and the precipitation column (**prcp**). The dataset also contains station metadata we can ignore. For our purposes, we only need the date and precipitation values.

```
# Keep only relevant columns and rename
prcp_df <- data.frame(
  date   = as.Date(jfk_prcp$date),
  precip = as.numeric(jfk_prcp$prcp)
)

summary(prcp_df)
```

```
      date                 precip
 Min.   :2000-01-01   Min.   :  0.000
```

```
 1st Qu.:2006-04-01   1st Qu.:  0.000
 Median :2012-07-01   Median :  0.000
 Mean   :2012-07-01   Mean   :  3.092
 3rd Qu.:2018-10-01   3rd Qu.:  1.000
 Max.   :2024-12-31   Max.   :204.500
```

Precipitation values in GHCN are typically stored in tenths of millimeters. We convert them to millimeters for easier interpretation.

Here, `prcp_df$precip` refers to the `precip` column inside the table `prcp_df`. The `$` works both ways: if the column already exists, `$` reads it; if you assign a value to a name that doesn't exist yet, `$` creates that column. We used `$` earlier to read a column; here we use it to create a new column called `precip_mm`.

```
# Create a new column
prcp_df$precip_mm <- prcp_df$precip / 10

# Show first rows
head(prcp_df)
```

```
        date precip precip_mm
1 2000-01-01    0.0      0.00
2 2000-01-02    0.0      0.00
3 2000-01-03    0.0      0.00
4 2000-01-04   15.7      1.57
5 2000-01-05    1.0      0.10
6 2000-01-06    0.0      0.00
```

A new column `precip_mm` is created by dividing the original `precip` values by 10, converting from tenths of millimeters to millimeters. This makes the precipitation values easier to interpret and compare with other datasets.

Now we merge the precipitation data with the groundwater dataset using the date as the common variable.

```
# Merge datasets by date
gw_prcp <- merge(
  merged_gw,
  prcp_df,
  by = "date",
  all = TRUE
)

head(gw_prcp)
```

```
        date site_no gwl precip precip_mm
1 2000-01-01    <NA>  NA    0.0      0.00
2 2000-01-02    <NA>  NA    0.0      0.00
3 2000-01-03    <NA>  NA    0.0      0.00
4 2000-01-04    <NA>  NA   15.7      1.57
5 2000-01-05    <NA>  NA    1.0      0.10
6 2000-01-06    <NA>  NA    0.0      0.00
```

You can think of `merge()` as matching rows between two tables using a shared column, similar to combining spreadsheets using a common ID or date.

2.4.5 When Dates Do Not Match

After merging the datasets, we notice that some rows contain values for precipitation but not for groundwater, and vice versa. Each dataset was collected independently and may have different measurement gaps or time coverage. These missing values are not errors; they reflect the reality of environmental data. When working with multiple datasets, it is common that one variable is available while another is not. Recognizing this helps us avoid misinterpreting incomplete records. The combined dataset now contains groundwater levels and precipitation for the same time period, and as before, some dates may have missing values in one dataset but not the other.

This situation extends the idea of missing data introduced earlier. Previously, we examined missing values within a single dataset. Here, missing values also arise from combining datasets with different time coverage. Before any analysis, it is important to decide how to handle these gaps. Depending on the goal, we may leave them as missing, interpolate short gaps, or focus only on periods where both variables are available.

In many analyses, we only use periods where both variables are available. This is called working with overlapping data. However, it is important to be transparent about how much data are being used and how much are missing. This transparency helps others understand the limitations of the analysis and any bias that may result from excluding certain periods.

```
# Check missing values after merge
colSums(is.na(gw_prcp))
```

```
     date   site_no       gwl    precip precip_mm
        0      5747      5747         0         0
```

This is expected when combining datasets from different sources.

2.4.6 Why Normalize?

Groundwater level and precipitation are measured in different units and can have very different ranges. For example, groundwater level may vary by a few feet, while precipitation values may range from zero to several millimeters in a single day. Because of these differences, plotting the two variables directly on the same axis can be misleading. One variable may appear flat simply because it has a smaller range. To compare patterns over time, we normalize the data, rescaling every value onto a common 0-to-1 ruler so that both variables sit on the same scale. Similar normalization workflows are widely used in sustainability analytics when comparing variables such as emissions, energy use, population, groundwater, precipitation, and climate indicators that use different units or scales.

Normalizing Variables for Comparison

Groundwater level and precipitation have different units and ranges, so we normalize them before plotting. Normalization rescales data so variables with different units can be compared on the same scale. This is especially important when plotting multiple variables together, as it allows us to see patterns and relationships without being misled by differences in magnitude. After normalization, all values fall between 0 and 1.

After normalization, the variables no longer retain their original units because the values have been rescaled between 0 and 1 for comparison.

We create a function so we don't have to repeat the same calculation multiple times. This function takes a vector of values and rescales it to the range of 0 to 1 based on the minimum and maximum values in that vector. By defining this function, we can easily apply it to both groundwater level and precipitation without having to write the same code twice. This makes our code cleaner and more efficient.

The main idea for the below code is simple:

- subtract the smallest value
- divide by the total range
- and rescale everything between 0 and 1

After normalization:

- 0 represents the smallest value
- 1 represents the largest value
- everything else falls somewhere in between

Here we build our own command. `function(x)` says "this command takes one input, which we'll call `x`"; `x` is a stand-in for whatever values we hand it later. `na.rm = TRUE` tells R to ignore missing values while finding the smallest and largest.

```
normalize <- function(x) {
  (x - min(x, na.rm = TRUE)) /
    (max(x, na.rm = TRUE) - min(x, na.rm = TRUE))
}
```

Defining a function for normalization helps simplify the code and makes it easier to reuse. Instead of repeating the same calculation multiple times, we write it once and apply it to different variables. This keeps the code easier to read and reuse. It also makes debugging easier, because any change to the calculation only needs to be made in one place.

The following line means:

- take the groundwater values,
- apply the `normalize()` function,
- and store the result in a new column called `gwl_norm`.

```
gw_prcp$gwl_norm <- normalize(gw_prcp$gwl)
```

Repeat for precipitation values, creating a new column called `precip_norm`.

```
gw_prcp$precip_norm <- normalize(gw_prcp$precip_mm)
```

With both variables now on the same 0-to-1 scale, we can plot them together and watch how each one responds over time.

```
plot(
  gw_prcp$date,
  gw_prcp$gwl_norm,
  type = "l",
  col = "blue",
  lwd = 3,
  lty = 1,
  xlab = "Date",
  ylab = "Normalized values",
  main = "Groundwater and Precipitation (Normalized)"
)

lines(
  gw_prcp$date,
  gw_prcp$precip_norm,
  col = "black",
  lwd = 1
```

```
)

legend(
  "topleft",
  legend = c("Groundwater (thick)", "Precipitation (thin)"),
  col = c("blue", "black"),
  lwd = c(3, 1),
  bty = "n"
)
```

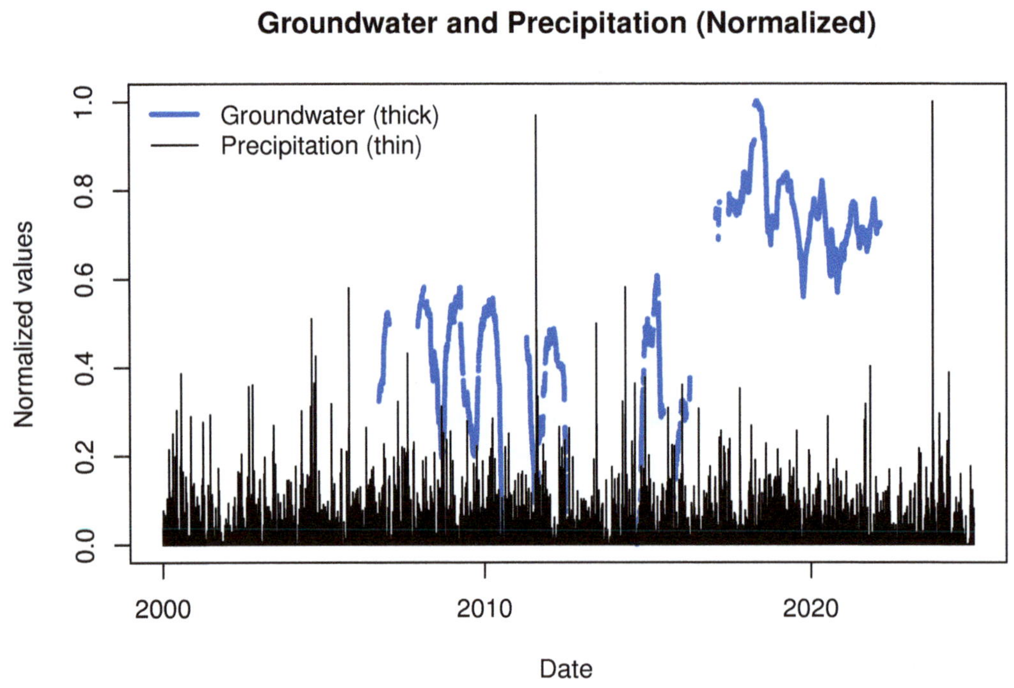

Figure 2.7: Normalized groundwater (thick blue line) and precipitation (thin black line) time series for comparison.

Groundwater responds more gradually than precipitation, reflecting slower subsurface processes (Figure 2.7). After normalization, both variables are scaled between 0 and 1, allowing us to compare their behavior over time. Precipitation (the thin black line) shows frequent short-term spikes, reflecting individual rainfall events. In contrast, groundwater (the thick blue line) changes more gradually, indicating a slower response of the subsurface system. This difference illustrates an important concept in geoscience: surface processes often vary quickly, while subsurface systems respond more slowly and smoothly. At this stage, we are not drawing conclusions, but beginning to explore how these variables might be related.

Although we are not performing a formal analysis yet, we can begin to ask simple questions. For example, do increases in precipitation tend to be followed by changes in groundwater? And if so, does that response happen immediately or after some delay? These types of questions will be explored in later chapters. For now, the key takeaway is that careful data handling allows us to combine and compare different datasets in a meaningful way. By normalizing the variables, we can compare their patterns more directly and explore potential relationships.

In this chapter, we prepared real-world geoscience data by cleaning, aligning, and combining multiple datasets. We also explored how groundwater and precipitation behave differently over time. To better understand these patterns and relationships, we now move to time series analysis, where we examine how data change through time and how variables may be connected.

Data-quality checklist

Before moving to analysis, make sure:

- Data are sorted by time
- Missing values are identified and understood
- Missing dates are checked and handled if needed
- Datasets are aligned correctly when merged
- Units and scales are understood
- Variables are normalized when comparing patterns

A clean dataset does not guarantee correct results, but it makes correct analysis possible.

2.5 What Makes Data Trustworthy

- Data handling is the foundation of reliable analysis.
- Inspect the structure before cleaning the data.
- Missing values and duplicates should always be checked.
- Time order matters and should be made explicit.
- Interpolation can fill small gaps, but it does not replace real observations.
- Clean data are not the final answer, but they make every later Earth and sustainability analysis more reliable.

Careful data preparation does not guarantee correct conclusions, but poor data preparation can easily lead to misleading results.

Where These Methods Lead

In this chapter you downloaded real measurements and cleaned them into a tidy table ready for analysis. Student research starts the same way, only at a larger scale.

Buchbinder and colleagues (2021) used R to download and clean about 238,000 groundwater records from 56 USGS monitoring stations across Long Island, New York. With the data in order, they compared groundwater levels before and during the COVID-19 stay-at-home period and found a clear change once people stayed home, one that reversed a rise lasting about fifteen years. Because groundwater levels reflect how much water people use, this points to a practical need: planning how a community protects its groundwater during a crisis, so that heavy use does not draw in salt water from the coast.

The careful download-and-clean work you practiced here is what makes a study like that possible.

2.6 From Analysis to Poster

A strong poster figure for this chapter could show:

- the raw groundwater series,
- the locations of missing values,
- the interpolated series,
- a short note explaining what was changed and why,
- and a simple map showing the location of the USGS monitoring station.

The next chapter introduces basic spatial mapping in R, which will allow us to visualize where environmental measurements are collected.

2.7 Exercises

1. Download a single USGS groundwater station and inspect its structure with `head()`, `str()`, and `summary()`.
2. Count the missing values in the groundwater record and make a simple plot of the missing points.
3. Create a date column if needed, then order the data by date.
4. Compare the raw groundwater series with a linearly interpolated version and describe one limitation of interpolation.
5. Find whether the table contains duplicate rows and explain whether they should be removed.
6. Download daily precipitation data from GHCN for a nearby station and merge it with the groundwater data by date.

2.7.1 References (Data and Further Reading)

1. USGS National Water Information System (NWIS). https://waterdata.usgs.gov/nwis

2. dataRetrieval package documentation. https://cran.r-project.org/package=dataRetrieval

3. GHCNr package documentation. https://cran.r-project.org/package=GHCNr

4. Global Historical Climatology Network (GHCN). https://www.ncei.noaa.gov/products/land-based-station/global-historical-climatology-network-daily

5. Wickham, H., Çetinkaya-Rundel, M., & Grolemund, G. *R for Data Science.* https://r4ds.hadley.nz

6. De Cicco, L. A., Hirsch, R. M., Lorenz, D., Watkins, W. D., & Johnson, M. (2024). *dataRetrieval: R packages for discovering and retrieving water data available from federal hydrologic web services [R package].* U.S. Geological Survey. https://doi.org/10.5066/P9X4L3GE

7. Menne, M. J., Durre, I., Korzeniewski, B., McNeill, S., Thomas, K., Yin, X., Anthony, S., Ray, R., Vose, R. S., Gleason, B. E., & Houston, T. G. (2012). *Global Historical Climatology Network-Daily (GHCN-Daily), Version 3.* NOAA National Climatic Data Center. https://doi.org/10.7289/V5D21VHZ

8. Buchbinder, B., Phan, E., Renner, F., Liebowitz, A., Achek, M., Marsellos, A. E., & Tsakiri, K. G. (2021). *An Approach to Study the COVID-19 Stay-at-Home Effect on Water Consumption in Long Island, NY Using Time Series Analysis of Groundwater Levels.* In 28th Conference on the Geology of Long Island and Metropolitan New York, April 2021.

9. Mighty, M., Neville, J., Bernhardt, L., Tsakiri, K. G., & Marsellos, A. E. (2026). *From Seasonal Stability to Event-Driven Dynamics: A Spectral Analysis of Groundwater Response to Winter Conditions Near JFK Airport, New York.* In 33rd Annual Conference on the Geology of Long Island and Metropolitan New York, April 2026.

10. Hope, J. H., Marsellos, A. E., & Tsakiri, K. G. (2024). *High-Frequency of Groundwater Level Fluctuations, Underground Erosion, Potential Sinkhole Occurrences Across Long Island, NY.* In 31st Conference on the Geology of Long Island and Metropolitan New York, April 2024.

Part II

Core Methods

Chapter 3

Spatial Analysis (GIS)

From Basic Maps to Elevation and DEMs in New York

In this chapter, we start with simple maps and gradually move toward real spatial analysis. Part A introduces basic spatial ideas and quick maps. Part B moves into coordinate systems, raster data, and simple GIS operations. Part C applies these ideas to Long Island and prepares the terrain analysis that Chapter 6 will reuse.

3.1 Introduction

In Chapter 2, we focused on preparing reliable datasets before analysis. In this chapter, we extend that same idea to spatial data, where location becomes part of the analysis. Spatial workflows require not only clean data, but also layers that line up correctly (the same coordinate system), a suitable level of map detail (spatial resolution), and a sensible real-world meaning. We explain each of these terms later in the chapter.

Spatial data simply means data that has a location (a place on Earth). A **Geographic Information System (GIS)** is a system used to work with spatial data. GIS helps us store, analyze, visualize, and interpret information connected to location. In simple terms, GIS helps us answer questions about "where".

Earthquakes, rivers, coastlines, elevation, climate, land use, flooding, and environmental vulnerability all depend on position on the Earth's surface. In earlier chapters, we worked with numbers and tables. Now we also include location with coordinates. In Earth science, "where" is often the most important part. For example:

- Earthquakes → where they happen shows plate boundaries
- Rainfall → where it falls shows climate patterns

So instead of just numbers, we use maps to understand patterns. Maps help us see patterns that are difficult to recognize in tables alone. For example, earthquake locations can help identify tectonic boundaries, temperature maps can show regional climate differences, and elevation maps can explain how landforms and flooding are related.

This kind of spatial workflow underlies real student research. Parag and colleagues (2025) used R to map vertical land motion from continuous GPS stations across Long Island, interpolating each station's long-term sinking rate into a single map of where the ground is slowly sinking.

We will start with simple maps and then move to elevation data (DEM), which helps us understand terrain.

> Why spatial visualization matters
>
> Maps help us see patterns that are hard to notice in tables.

> What you will be able to do
>
> By the end of this chapter, you will be able to:
>
> - recognize what a shapefile is and what it contains
> - open spatial data in R and inspect its structure
> - save spatial data for reuse
> - check whether spatial layers use the same coordinate system
> - transform a layer to a new coordinate system when needed
> - combine multiple layers on a single map
> - perform basic spatial operations such as clipping data to a study area
> - connect terrain and spatial patterns to environmental and sustainability questions

There are two main types of spatial data: vectors and rasters.

3.1.1 Vector data

Vector data is used when we want to draw clear boundaries, such as borders, roads, or coastlines. It is made up of three main types of features:

A simple way to think about vector data:

- point → a single location (like a city)
- line → something that connects places (like a road)
- polygon → an area (like a state)

Use vector data when you need **clear and exact boundaries**, such as borders, roads, or coastlines.

> A simple way to remember it
>
> Vector data is like drawing precise objects on a map. Each feature has a clear shape, a location, and often an attached table of facts about it (called attributes, such as a name or population). You can zoom in without seeing pixels because vector boundaries stay smooth.

3.1.2 Raster data

Raster data is like a photo made of pixels. Each square (pixel) stores a value, for example:

- elevation (height)
- temperature
- rainfall

Key idea:

- **Vector** = shapes
- **Raster** = grid of values

The size of each square controls how detailed the map is.

- small squares → finer, more detail
- large squares → coarser, less detail

This square size is called the **spatial resolution**.

Cells with no data are stored as missing values (NA).

> Resolution matters
>
> A raster that is too coarse may hide important terrain details. In coastal studies, this can mean missing narrow, low-lying areas that are critical for flooding. Each pixel is like a square that averages everything inside it into a single value.

> Tip
>
> **Vector draws boundaries, raster fills space with values.**

Raster data becomes most useful when it is connected to a specific study area, which we define next.

If you are new to spatial data, start with Part A and move forward step by step.

3.2 Part A: Basic Spatial

3.2.1 Simple Maps in R

We will use the same idea from Chapter 1: loading a package and using its functions. Here, we will use them to make maps. To quickly remember what we did, here is another example of drawing Italy in Europe:

```
library(maps) # Load the maps package for drawing maps
maps::map("world", regions = "italy") # Draw the outline of Italy
title("Italy") # Add a title to the map
```

Figure 3.1: Simple map of Italy using the maps package.

This simple outline map is useful for introducing spatial visualization with the `maps` package (Figure 3.1).

Here is another example showing how to plot a selected group of six states on one figure, with a major city represented as a point in each state. Instead of including one state name, we combined multiple state names into a vector (a group of values) using the `c()` function. We can later use this vector for selection and filtering.

```
states_to_plot <- c("new york", "florida", "texas",
                    "california", "illinois", "ohio")
maps::map("state", regions = states_to_plot,
          fill = TRUE, col = "lightblue")
points(c(-73.75, -82.55, -97.74, -119.42, -89.38, -82.99),
       c(42.65, 27.99, 30.27, 34.05, 41.88, 39.96),
       pch = 19) # Add points for major cities
# Add a title to the map
```

```
title("Six U.S. States with Major Cities")
```

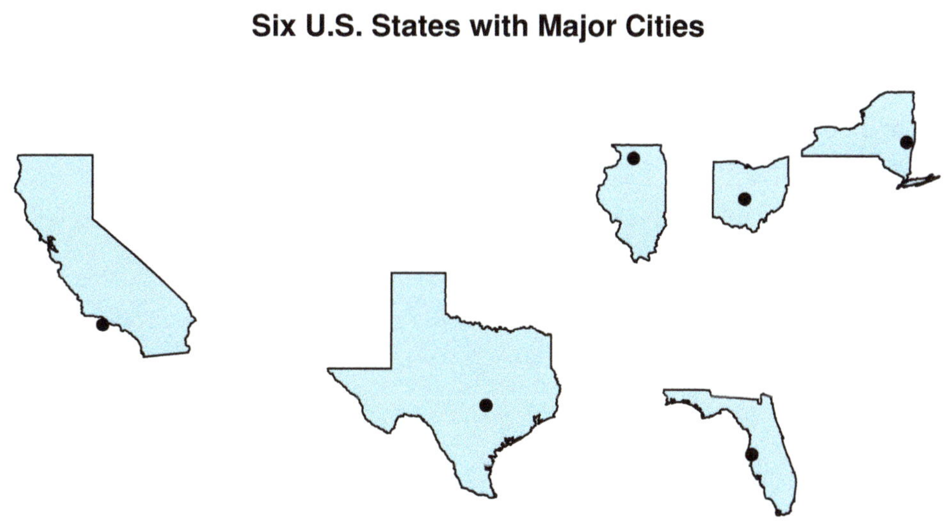

Figure 3.2: Map of six U.S. states with major cities marked.

Multiple states and their major cities can be combined into a single visualization, illustrating how selected polygons (here, the state shapes) can be added to a map (Figure 3.2).

3.2.2 Plotting Maps

The first step in spatial analysis is often a simple map. A basic map helps us see the shape of the study area before adding more detail.

Add a second layer

We added major cities in the previous map. Here is another quick map showing volcano locations around the world. This is a good example of how to plot many points from a `CSV` file on a map. The first map shows volcanoes globally, while the second zooms into Greece to reveal regional clustering more clearly. This illustrates how spatial data can be visualized at different scales to reveal patterns that may not be visible in a global view. The volcano locations come from the plotly sample-data collection (the `volcano_db.csv` file), originally based on the Smithsonian Institution Global Volcanism Program.

```r
# Load volcano data from an online source
volc <- read.csv(
  "https://raw.githubusercontent.com/plotly/datasets/master/volcano_db.csv"
)
# Draw the world map as a base layer
map("world", fill = TRUE, col = "lightgray")
# Add points for volcano locations
points(volc$Longitude, volc$Latitude,
       col = "darkred",
       pch = 19, cex = 0.5)

title("Global Volcano Centers")
```

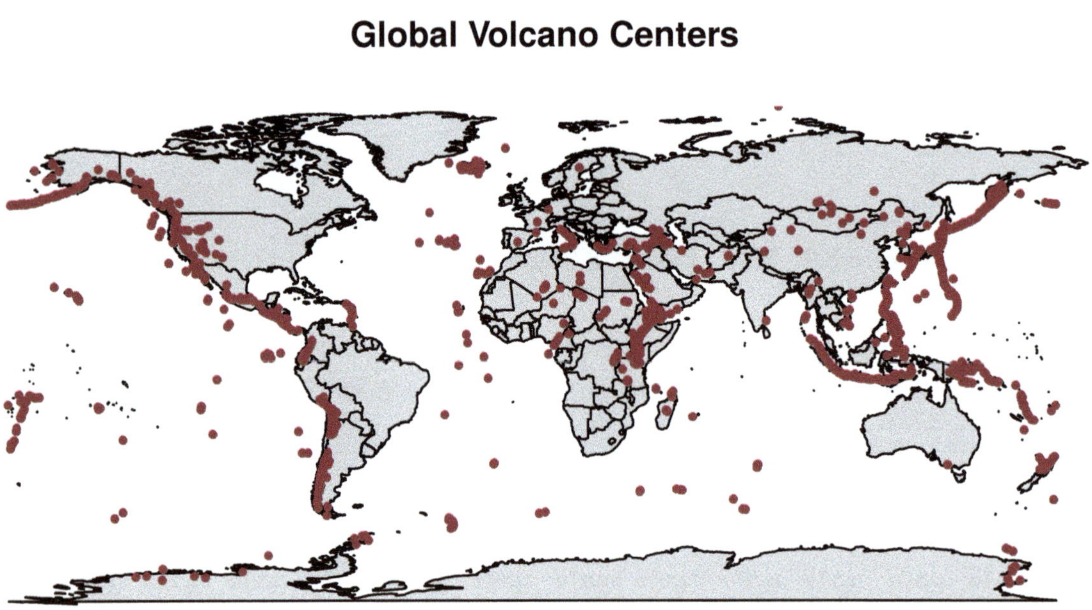

Figure 3.3: Global volcano locations plotted on a world map.

The global distribution of volcanoes reveals patterns associated with tectonic boundaries (Figure 3.3). We can zoom into Greece to see the pattern more clearly. Greece has many volcanoes, and zooming in allows us to see their distribution in relation to the country's geography.

```
# Zoom into Greece using world map limits
map("world",
    xlim = c(19, 30),    # longitude range
    ylim = c(34, 42),    # latitude range
    fill = TRUE,
    col = "lightgray")
points(volc$Longitude, volc$Latitude,
       col = "darkred", pch = 17, cex = 1.2)
title("Zoomed Map of Greece")
```

Figure 3.4: Volcano locations in Greece plotted on a zoomed world map.

Zooming reveals volcanic clustering in Greece that is less apparent at the global scale (Figure 3.4). These clusters are associated with active tectonic processes in the region.

Built-in map data

The **maps** package already contains map data, so we can draw maps without downloading any additional files.

```
maps::map("world") # map the entire world
maps::map("world", regions = "france") # map France
US_states <- maps::state.fips # US state FIPS codes and names
US_counties <- maps::county.fips # US county FIPS codes and names
```

The map (Figure 3.4) shows how additional layers can provide context to the map. Maps are easier to read when they are simple. Start with the boundary, then add one layer at a time.

Build maps gradually

A clean map is often more effective than a crowded one. Add one layer, check it, and then add the next.

3.2.3 Working with Attributes

Exploring the Data Behind the Map

Spatial objects usually have two parts:

- the **geometry**, which tells us where the feature is (coordinates), and
- the **attributes**, which describe what the feature is (a table of information).

For example, a state polygon may contain attribute fields such as name, population, or code. The geometry tells us the shape and location (e.g. point, line, or polygon features), while the attributes tell us more about what that shape represents.

Each point on the map represents a volcano, but each volcano also has information attached to it. That information is stored in the table we loaded. Let's take a quick look at the data. The `head()` function shows us the first few rows of the volcano data, which includes columns for the volcano's name, location (latitude and longitude), elevation, and other attributes. This table is linked to the points on the map, so we can use it to understand more about each volcano.

```
head(volc)
```

```
    Number   Volcano.Name    Country        Region Latitude Longitude Elev
1 0803-001            Abu      Japan  Honshu-Japan   34.500   131.600  571
2 1505-096     Acamarachi      Chile       Chile-N  -23.300   -67.620 6046
```

```
3 1402-08=      Acatenango     Guatemala     Guatemala  14.501   -90.876 3976
4 0103-004 Acigol-Nevsehir        Turkey        Turkey  38.570    34.520 1689
5 1201-04-           Adams United States US-Washington  46.206  -121.490 3742
6 0803-17=         Adatara         Japan  Honshu-Japan  37.620   140.280 1718
            Type           Status Last.Known
1 Shield volcano         Holocene    Unknown
2  Stratovolcano         Holocene    Unknown
3  Stratovolcano       Historical         D1
4           Maar         Holocene          U
5  Stratovolcano Tephrochronology         D6
6  Stratovolcano       Historical         D1
```

The table includes:

- name of the volcano
- country
- latitude and longitude (location)
- elevation
- type of volcano

So the map shows *where*, and the table tells us *what*. In a GIS workflow, the geometry and attributes are stored together as a spatial object. This is a key idea in GIS: spatial features combine location with descriptive information.

```
table(volc$Type)
```

```
           Caldera         Cinder cone        Cinder cones     Complex volcano
                84                  76                   4                  53
  Compound volcano                Cone         Crater rows    Explosion crater
                11                   3                   8                   4
      Fissure vent      Fissure vents      Fumarole field  Hydrothermal field
                19                   1                   4                   2
         Lava cone           Lava dome          Lava domes                Maar
                 4                  37                   3                  19
             Maars         Mud volcano        Not Volcanic         Pumice cone
                 1                   1                   1                   3
  Pyroclastic cone   Pyroclastic cones  Pyroclastic shield         Scoria cone
                36                   8                   6                   7
      Scoria cones      Shield volcano   Shield volcanoes      Somma volcano
                 3                 169                   2                   5
     Stratovolcano     Stratovolcanoes  Subglacial volcano   Submarine volcano
               704                   9                   5                 142
 Submarine volcano? Submarine volcanoes           Tuff cone          Tuff rings
```

```
               4                 5                13                 1
         Unknown  Volcanic complex     Volcanic field
              17                 2                95
```

It counts how many volcanoes belong to each type. So now we are not just plotting points; we are starting to analyze the data behind the map. For example, how many stratovolcanoes are there? Are there any duplicate names (e.g. cinder cone and cinder cones, which differ only by an s)? We can filter the data to find out or plot only the stratovolcanoes on the map using a variable `strato`:

```
strato <- volc[volc$Type == "Stratovolcano", ]
nrow(strato)
```

```
[1] 704
```

In many GIS workflows, we combine spatial layers with external tables. For example, a country polygon layer may be combined with population, climate, or economic data stored in a CSV file. The polygon provides the shape and location, while the table provides additional information that can be mapped and analyzed together. Later in the book, we will combine global country boundaries with environmental and sustainability datasets.

So far, we have created maps by placing points on top of a background and explored the data behind them. This is an important first step because it helps us connect location with real-world information. However, these maps are still mainly visualizations. In a real GIS workflow, the map features, their coordinates, and their attribute tables are stored together as spatial objects. These spatial datasets can then be saved, reopened, projected, combined with other layers, and analyzed. One of the most common ways to store vector spatial data is the shapefile format, which we introduce next.

We are still working with simple maps, but these same ideas will soon allow us to analyze real terrain, flooding, and Earth processes.

In Part A, we used the **`maps`** package for quick visualization. In Part B, we move to two new packages: **`sf`** (for vector data such as points, lines, and polygons) and **`terra`** (for raster grids such as elevation). They store geometry, attributes, and coordinate systems together for GIS analysis. In other words, we move toward working with spatial data as GIS objects. These tools allow us to combine layers correctly and measure distances and areas accurately.

3.3 Part B: Core GIS Workflow

In this part, we move from simple map visualization to working with real GIS datasets and spatial objects in R. The goal is not to memorize every GIS file format or function. The goal is to understand how location, attributes, and spatial relationships work together. To begin working with real GIS data, we first need to understand how spatial layers are stored.

3.3.1 What is a shapefile?

Earlier in this chapter, we learned that spatial data usually contains:

- geometry (the shape and coordinates)
- attributes (the information attached to the object)

A shapefile is one of the most common GIS file formats used to store geometry and attributes together as a spatial layer.

When a shapefile is opened in R with the `sf` package, it becomes an `sf` spatial object that we can analyze and map. Shapefiles are commonly used for vector data such as:

- points
- lines
- polygons

For example:

- a point shapefile may store earthquake locations,
- a line shapefile may store roads or rivers,
- a polygon shapefile may store counties or states.

> **i A shapefile is not just one file**
>
> A shapefile is actually a group of files that work together in the same folder.

Common shapefile components include:

File	Purpose
`.shp`	stores the geometry (shapes)
`.dbf`	stores the attribute table
`.shx`	stores the shape index
`.prj`	stores the coordinate system

For example, a New York boundary shapefile may appear as:

```
new_york.shp
new_york.dbf
new_york.shx
new_york.prj
```

For most analyses, you do not need to memorize the file extensions. The important idea is that a GIS layer contains both geometry and attributes in more than one file.

> ⚠ Keep shapefile components together
>
> If one of the shapefile components is deleted or moved, the layer may not open correctly in GIS software or in R.

You can think of a shapefile as a table connected to map features. Each row represents one spatial feature, and each feature has both a shape on the map and information stored in columns.

> ℹ Geometry + attributes = GIS
>
> A GIS layer is not just a picture of a map. It also contains coordinates, shapes, attribute tables, and coordinate system information that can be analyzed.

3.3.2 Reading and saving spatial data

Spatial layers can be opened and saved in R just like CSV files.

The `sf` package uses:

- `st_read()` → open spatial data
- `st_write()` → save spatial data

The following example demonstrates the basic syntax used to open and save spatial files in R. You do not need to run this example unless you already have a shapefile stored on your computer.

```
library(sf)

# Open a shapefile
ny <- st_read("data/new_york.shp")

# Save a shapefile
st_write(ny, "output/new_york.shp")
```

Modern GIS workflows also commonly use GeoPackage (.gpkg) files, which store spatial data in a single file instead of multiple shapefile components.

```
# Save as GeoPackage
st_write(ny, "output/new_york.gpkg")
```

> GeoPackage vs shapefile
>
> A **shapefile** uses multiple files (.shp, .dbf, .shx, .prj), while a **GeoPackage** stores everything inside one .gpkg file.

In this part, we move from simple plotting to working with spatial data as true objects in R. We will use the `sf` package for vector data and the `terra` package for raster data. These tools allow us to store location and attributes together, so we can analyze, measure, and combine spatial data correctly. A key step in this process is understanding coordinate systems, which ensure that different layers align and measurements are accurate.

3.3.3 Coordinate Systems

At first, coordinate systems can feel difficult to visualize. The important idea is simple:

A coordinate system tells the computer how to take a place on the round Earth and show it on a flat map, the same problem you face when you try to flatten an orange peel without tearing it.

3.3.4 What is a coordinate system?

A coordinate system tells us how to describe a location on Earth. Think of it like an address system:

- latitude and longitude → global address
- projected coordinates → local measuring system

> Degrees or meters
>
> Locate with latitude/longitude, measure with projected coordinates.

Common coordinate system names

You may encounter code-like names such as NAD83, WGS84, and UTM when working with spatial data. They are simply the names of different coordinate systems, like brand names for different ways of describing locations.

- NAD83 and WGS84 are commonly used geographic coordinate systems that use latitude and longitude to describe locations on Earth.
- UTM is a projected coordinate system that uses meters instead of degrees, making it more suitable for measuring distance and area.

For now, do not worry about the technical differences between NAD83 and WGS84. The important idea is that geographic coordinate systems are mainly used for locating features, while projected coordinate systems are used for accurate measurements.

We use different coordinate systems depending on whether we want to locate features globally or measure distances and areas accurately within a region. We choose the system based on what we want to do. In this chapter, we will use projected coordinates when we need accurate measurements for our study area in New York.

Why it matters

A map can look correct even when its layers do not truly match. That often happens when layers use different coordinate systems. Before combining layers, always make sure they use the same coordinate system.

A simple way to think about this is:

- **Longitude and latitude** describe positions on a curved globe.
- **Projected coordinates** flatten the Earth into a map so that distances and areas can be measured more consistently.

Longitude and latitude divide the curved Earth into sections that help us locate places on the globe, similar to a grid wrapped around a sphere. However, the distances inside those sections are not constant everywhere on Earth. Because of this, maps that use degrees are good for location, but not ideal for accurate measurement. Projected coordinate systems convert the curved Earth into a flat map where distances and areas can be measured more consistently using units such as meters.

For a first spatial workflow, the key idea is not the math. The key idea is that layers must "speak" the same spatial language before they can be compared.

> Thinking Like a Spatial Analyst
>
> Before combining spatial layers, ask:
>
> - Are the coordinate systems the same?
> - Are the spatial resolutions appropriate?

- Does the study area make scientific sense?
- Could projection choice affect measurements?
- Is the map showing visualization or analysis?

Good GIS analysis is not only about making maps. It is about making spatial relationships scientifically meaningful.

3.3.5 Example of a projection change

In R, spatial objects are often handled with the **sf** package for vector data and the **terra** package for raster data. When we read in a map, it may be in a geographic coordinate system (longitude and latitude). To do spatial analysis, we often need to project the data into a projected coordinate system that uses meters or feet. This allows us to measure distances and areas accurately.

We will also use the **tigris** package to get the New York State county boundaries as a spatial object. The example below shows how to download boundary data, check its coordinate system, and then project it to a new coordinate system suitable for local analysis. County boundaries are available at multiple spatial resolutions, and we will use a resolution of 500k to keep the boundaries detailed enough for our study area while still making the map fast to display. The `cb = TRUE` argument downloads simplified county boundaries. These simplified boundaries are easier and faster to display on maps than highly detailed boundaries.

```r
library(sf) # Work with spatial (vector) data
library(tigris) # Download U.S. boundary data
# Download U.S. counties for New York State (500k scale)
ny_state <- counties(state = "NY", cb = TRUE, resolution = "500k", year = 2024)
```

We now convert (transform) the county layer into a projected coordinate system called UTM Zone 18N. Transforming means recalculating every coordinate so the same places are described in a system suited for measuring distances and areas. ("UTM" is just the name of a popular flat-map measuring system, and "Zone 18N" is the slice of the Earth that covers New York.) Each such system has a short ID number called an EPSG code; the code for UTM Zone 18N is 32618. We use this system because it preserves distances and areas accurately for our study area in New York, which is essential for accurate Earth measurements and spatial analysis.

```r
# Transform to projected system
ny_state_utm <- st_transform(ny_state, 32618)
```

So EPSG:32618 is simply the ID number for the UTM Zone 18N coordinate system, which is commonly used in the northeastern United States. To find the right EPSG code for a different city or country, visit https://epsg.io and search by place name or coordinate system name.

We use **st_geometry()** to plot only the spatial geometry (the shapes) without plotting the attribute table.

```
# Let's plot the projected data
plot(st_geometry(ny_state_utm),
     col = "#e5f5e0",
     border = "#31a354",
     main = "New York State in a Projected Coordinate System",
     cex.main = 0.6)
```

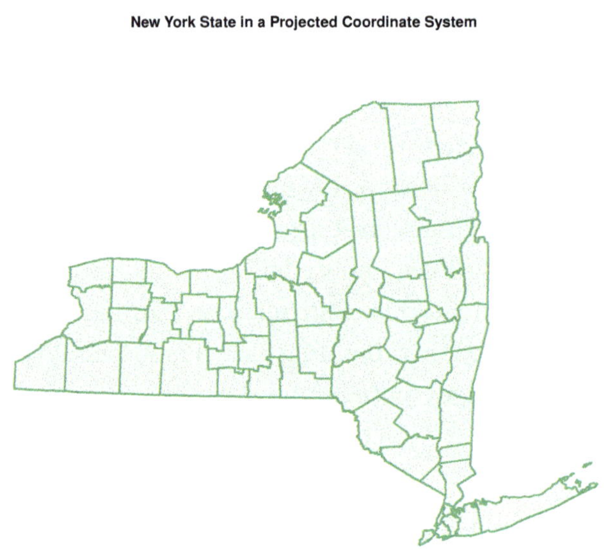

Figure 3.5: New York State transformed into a projected coordinate system for spatial analysis.

The transformation to a projected coordinate system allows distances to be measured in meters rather than degrees (Figure 3.5). This is why projected coordinate systems are preferred for distance, area, and terrain analysis. Later in the book, projected coordinates will allow us to estimate flooded area and compare spatial measurements correctly.

3.3.6 Selecting Features from a Spatial Layer

The New York State county layer contains multiple county polygons. We will focus on Long Island counties. To select multiple counties at once, we use `%in%`, which checks whether a value belongs to a list. For example, `ny_state$NAME %in% c("Kings", "Queens", "Nassau", "Suffolk")` checks whether each county name stored in `ny_state$NAME` matches one of the four Long Island county names combined into a vector using `c()`.

As shown earlier, we will use `%in%` to select the Long Island counties.

```
li <- ny_state[ny_state$NAME %in%
                 c("Kings", "Queens",
                   "Nassau", "Suffolk"), ]
```

This uses the same `%in%` operator introduced in Chapter 2. We are selecting only the counties whose names match the list.

Also, we will transform to UTM Zone 18N (`EPSG:32618`), the same projected system we used for the state layer.

```
# Print the full coordinate-system description for the layer
st_crs(li)
```

```
Coordinate Reference System:
  User input: NAD83
  wkt:
GEOGCRS["NAD83",
    DATUM["North American Datum 1983",
        ELLIPSOID["GRS 1980",6378137,298.257222101,
            LENGTHUNIT["metre",1]]],
    PRIMEM["Greenwich",0,
        ANGLEUNIT["degree",0.0174532925199433]],
    CS[ellipsoidal,2],
        AXIS["latitude",north,
            ORDER[1],
            ANGLEUNIT["degree",0.0174532925199433]],
        AXIS["longitude",east,
            ORDER[2],
            ANGLEUNIT["degree",0.0174532925199433]],
    ID["EPSG",4269]]
```

```
# Check units
st_crs(li)$units_gdal
```

```
[1] "degree"
```

```
# Transform to projected system
li_utm <- st_transform(li, 32618)
# Check units
st_crs(li_utm)$units_gdal
```

```
[1] "metre"
```

The `st_crs()` function checks which coordinate system a layer currently uses. It prints a lot of text, so do not worry about reading all of it.

Focus first on these three keywords in what it prints:

- `GEOGCRS` → it is a geographic (latitude/longitude) coordinate system
- `EPSG:4269` → the ID number of that coordinate system
- `degree` → the coordinates are stored as latitude and longitude

Later, we will transform the data into a projected coordinate system that uses meters instead of degrees.

Notice that geographic coordinate systems use degrees, while projected coordinate systems often use meters. Degrees are useful for locating positions globally, while meters are more useful for measuring distance, area, and terrain.

In this chapter, we use geographic coordinates (longitude and latitude, such as NAD83 / EPSG:4269 in this example) to retrieve or locate data, and projected coordinates (UTM) when we need accurate measurements and spatial analysis.

Plot Long Island Counties

We will now plot the Long Island spatial layer after transforming it to a projected coordinate system (`li_utm`).

```
# Let's plot the projected data
plot(st_geometry(li_utm),
     col = "#e5f5e0",
     border = "#31a354",
     main = "Long Island, NY in a Projected Coordinate System",
     cex.main = 0.6)
```

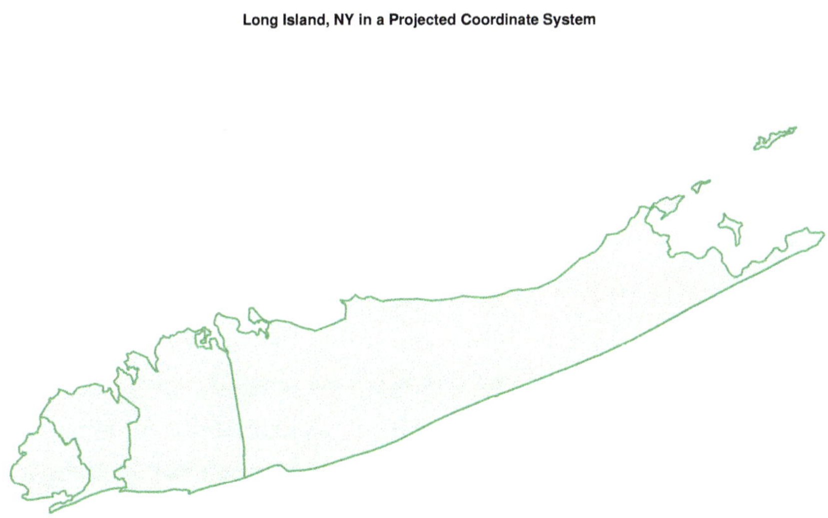

Figure 3.6: Long Island counties transformed into a projected coordinate system for spatial analysis.

The Long Island counties are now shown in the same projected coordinate system used later for terrain and coastal analysis (Figure 3.6). Using projected coordinates keeps spatial measurements consistent across the workflow.

Calculate area of Long Island counties

Now that we have the Long Island counties in a projected coordinate system, we can calculate their area using the `st_area()` function from the `sf` package. This function computes the area of each polygon feature in the spatial object.

```
# Calculate area
area_km2 <- st_area(li_utm)

# Convert units from m^2 to km^2
units(area_km2) <- "km^2"

# Show result
data.frame(
  county = li_utm$NAME,
  area_km2 = round(area_km2, 1) # round to one decimal place
```

```
)
```

```
   county      area_km2
1   Kings  205.3 [km^2]
2  Nassau  814.4 [km^2]
3 Suffolk 2978.6 [km^2]
4  Queens  318.7 [km^2]
```

The NAME column comes from the attribute table inside the spatial object. We combine the county names with the calculated areas into a regular table for display.

Because the layer is projected into meters, R can now calculate area using real-world units. Each polygon (county) has its own measured area. This would not work correctly using longitude and latitude coordinates because degrees are not true distance units.

3.3.7 Selecting Columns from an sf Object

We now apply the idea of attributes to real spatial data using the sf package. An sf object contains both geometry (shape) and attributes (a table of information). We can inspect and select these attributes using standard data frame operations.

```
# Look at the first few attribute columns
head(ny_state)
```

```
Simple feature collection with 6 features and 12 fields
Geometry type: MULTIPOLYGON
Dimension:     XY
Bounding box:  xmin: -79.13689 ymin: 40.78536 xmax: -73.29361 ymax: 44.54684
Geodetic CRS:  NAD83
   STATEFP COUNTYFP COUNTYNS        GEOIDFQ GEOID        NAME
41      36      005 00974101 0500000US36005 36005       Bronx
43      36      119 00974157 0500000US36119 36119 Westchester
44      36      053 00974125 0500000US36053 36053     Madison
47      36      029 00974113 0500000US36029 36029        Erie
49      36      091 00974143 0500000US36091 36091    Saratoga
90      36      031 00974114 0500000US36031 36031       Essex
             NAMELSAD STUSPS STATE_NAME LSAD      ALAND    AWATER
41       Bronx County     NY   New York   06  109235672  39353304
43 Westchester County     NY   New York   06 1115811940 179391718
44     Madison County     NY   New York   06 1696054034  16691322
47        Erie County     NY   New York   06 2700743425 476876396
```

```
49    Saratoga County     NY   New York   06 2097812305  87488004
90       Essex County     NY   New York   06 4646753184 316350563
                         geometry
41 MULTIPOLYGON (((-73.77242 4...
43 MULTIPOLYGON (((-73.77237 4...
44 MULTIPOLYGON (((-75.99243 4...
47 MULTIPOLYGON (((-79.13689 4...
49 MULTIPOLYGON (((-74.1601 43...
90 MULTIPOLYGON (((-74.33683 4...
```

```
# Example - select only columns we need
ny_state_small <- ny_state[, c("NAME", "STUSPS", "geometry")]
head(ny_state_small)
```

```
Simple feature collection with 6 features and 2 fields
Geometry type: MULTIPOLYGON
Dimension:     XY
Bounding box:  xmin: -79.13689 ymin: 40.78536 xmax: -73.29361 ymax: 44.54684
Geodetic CRS:  NAD83
         NAME STUSPS                       geometry
41      Bronx     NY MULTIPOLYGON (((-73.77242 4...
43 Westchester    NY MULTIPOLYGON (((-73.77237 4...
44    Madison     NY MULTIPOLYGON (((-75.99243 4...
47       Erie     NY MULTIPOLYGON (((-79.13689 4...
49   Saratoga     NY MULTIPOLYGON (((-74.1601 43...
90      Essex     NY MULTIPOLYGON (((-74.33683 4...
```

This linkage between geometry and attributes is a core idea in GIS. The map tells us where, and the table tells us more about what is there. The `head(ny_state)` command shows a multipolygon geometry together with multiple attribute columns describing each county. We then keep only three columns and save the result as a new spatial object called `ny_state_small` in the R environment memory, not on the hard drive. Besides changing the number of columns, we can change the way that coordinates are shown.

When we move from geographic coordinates to projected coordinates, we prepare the data for area, distance, and slope calculations (Figure 3.5). The map may look similar, but the underlying coordinate system has changed to one that is better suited for accurate local measurements and spatial analysis.

⚠ Matching coordinate systems

If coordinate systems do not match, spatial measurements and layer alignment may become misleading.

3.3.8 Raster Data

Raster data is especially important in Earth and sustainability science because it naturally represents continuous environmental phenomena.

Examples include:

- elevation,
- slope,
- temperature,
- rainfall,
- vegetation,
- flood depth.

As a quick reminder, a raster is a grid where each cell stores one value linked to a geographic location.

3.3.9 Choosing a Study Area

In later chapters, we will use elevation data to identify flood-prone areas in Long Island. For now, we focus on preparing the terrain correctly. Before we download a DEM, we need to decide where we want to study. A study area can come from a place name, a set of coordinates, or a boundary drawn in another program such as Google Earth.

This step is important because it tells the computer what part of the Earth we want to analyze. In GIS work, this area is often called the region of interest, or ROI.

Once the ROI is defined, we can use it to clip elevation data so we only keep the part we actually need.

Begin with a clear boundary

A good GIS workflow starts with a clear study area. When the boundary is defined first, the rest of the analysis becomes easier to understand and easier to reproduce.

Why raster cells matter

- Each raster cell answers one question at one place. (e.g. one pixel equals one elevation)
- Because each raster cell has a real size on the ground, we can later estimate area by counting cells.

Space and time together

Some Earth datasets contain many raster layers through time, such as monthly temperature maps or yearly ice thickness measurements. In those cases, we analyze both space and time together.

3.3.10 Create a Digital Elevation Model

A digital elevation model (DEM) is a map of elevation (height). Each pixel tells you:
"How high is the land here?"
Example:

- mountains → high values
- coast → low values

The zoom level (`z`) controls how detailed the elevation raster will be.

- small z → less detail (faster)
- large z → more detail (slower)

DEM for the Northeastern United States

Let's begin with a broad regional view of elevation across the northeastern United States, using New York State as our region of interest (ROI). We draw the elevation map first, and then we add the state boundary on top using `add = TRUE`.

For a state-wide overview, we can request a moderate-resolution elevation surface (with zoom `z = 6`). The example below uses a relatively coarse resolution to keep the workflow fast. For publication or detailed analysis, higher-resolution data should be used.

We will use the `elevatr` package to download elevation data. The `get_elev_raster()` function allows us to specify the area of interest and the zoom level (which controls resolution). The `clip = "locations"` argument ensures that we only keep elevation data within the ROI area. The `clamp` function from `terra` lets us set an upper and lower limit so we only display elevations in a chosen range (for example, just low coastal land or just high mountains). Anything outside the range is simply not shown. This is useful for highlighting features of interest within the study area.

```
# Step 1: Load libraries
library(elevatr) # for downloading elevation data
library(tigris)  # for downloading boundary data
library(terra)   # for working with raster data
library(sf)      # for working with vector data
# Step 2: Get state boundaries
us_states <- tigris::states(cb = TRUE, resolution = "500k", year = 2024)
# Keep New York
ny_state <- us_states[us_states$NAME == "New York", ]
# Step 3: Download elevation data
# Download an elevation raster around the area of New York State
# A moderate zoom level 6 keeps the mapping process quick and easy
NE_dem <- elevatr::get_elev_raster(
  # specify the area of interest: New York State boundary
  locations = ny_state,
  z = 6 # resolution from 1-14, higher is more detailed
)
# Step 4: Plot the elevation data
plot(
  clamp(NE_dem, lower = 0),
  main = "Elevation for Northeastern region",
  cex.main = 0.6,
  col = terrain.colors(30)
)
 # Add state boundaries on top
plot(
  st_geometry(ny_state),
  add = TRUE,
  border = "black",
  lwd = 1
)
grid()
```

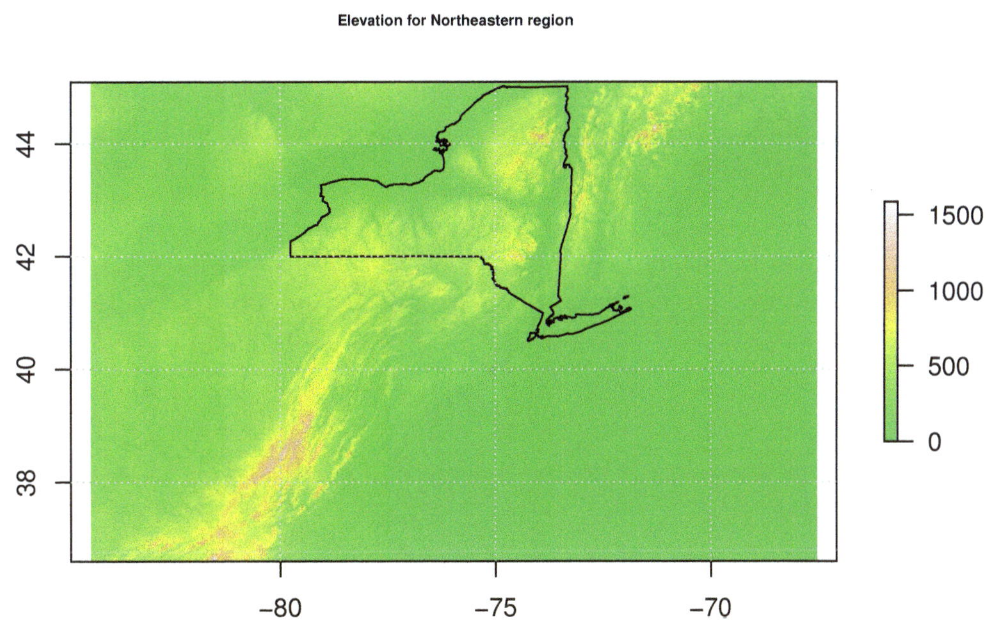

Figure 3.7: Elevation across the northeastern United States with the New York State boundary overlay.

The Adirondack Mountains in northern New York (Figure 3.7) show the highest elevations, while coastal regions remain relatively low. This regional DEM shows broad terrain patterns across the northeastern United States and provides context for later terrain and coastal analysis.

The DEM workflow follows four main steps:

1. load the spatial packages,
2. define the study area,
3. download the elevation data,
4. visualize and interpret the terrain.

Understanding terrain is also important for urban planning, infrastructure design, ecosystem management, and climate resilience.

> **i** Why DEMs are useful
>
> Elevation strongly influences where water can flow or accumulate, making DEMs essential for flood analysis, environmental planning, and sustainability studies. Low areas collect water, high areas stay dry. This helps us identify areas that may be vulnerable to flooding.

DEM for New York State

Instead of repeating the previous steps, which is not required because everything is saved in the environment (see top-right environment tab in RStudio), we can reuse the previous data and just apply the `clamp()` function to filter the elevation values, and the `clip = "locations"` to crop any information outside New York State.

We can constrain elevation ranges to focus on specific terrain features within our area of interest, here only New York State. For example, we can show only elevations above sea level that highlight terrestrial areas while we exclude below-sea-level coastal and underwater regions. Notice that we do not need to reload the libraries because they are already available in the current R session.

```
# Get elevation data
ny_dem <- get_elev_raster(
  locations = ny_state,
  z = 6, # resolution from 1-14, higher is more detailed
  # trim out any other information besides New York State
  clip = "locations"
)
# Plot elevations greater than 0 meters
plot(
  clamp(ny_dem, lower = 0),
  main = "Elevation (>0 m) for New York State",
  cex.main = 0.6,
  col = terrain.colors(30)
)
grid()
```

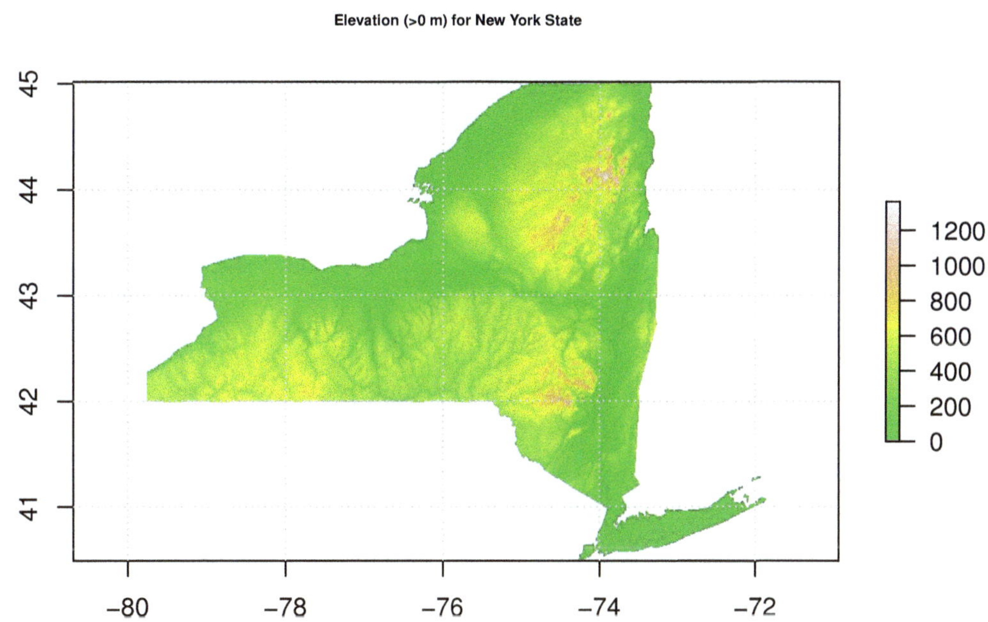

Figure 3.8: Elevation for New York State with values below sea level removed.

We can also constrain the elevation range within the DEM using the `clamp()` function from the `terra` package to show areas up to 600 m and eliminate regions lower than 0 m. This is useful when we want to focus on specific elevation ranges, such as mountainous areas or coastal zones. The map (Figure 3.9) displays the terrain of New York State while excluding areas below sea level and above 600 m, which can help us focus on the most relevant terrain features of the region for certain analyses.

```
plot(
  clamp(ny_dem, upper = 600, lower = 0),
  main = "Elevation for New York State",
  cex.main = 0.6,
  col = terrain.colors(30)
)
grid()
```

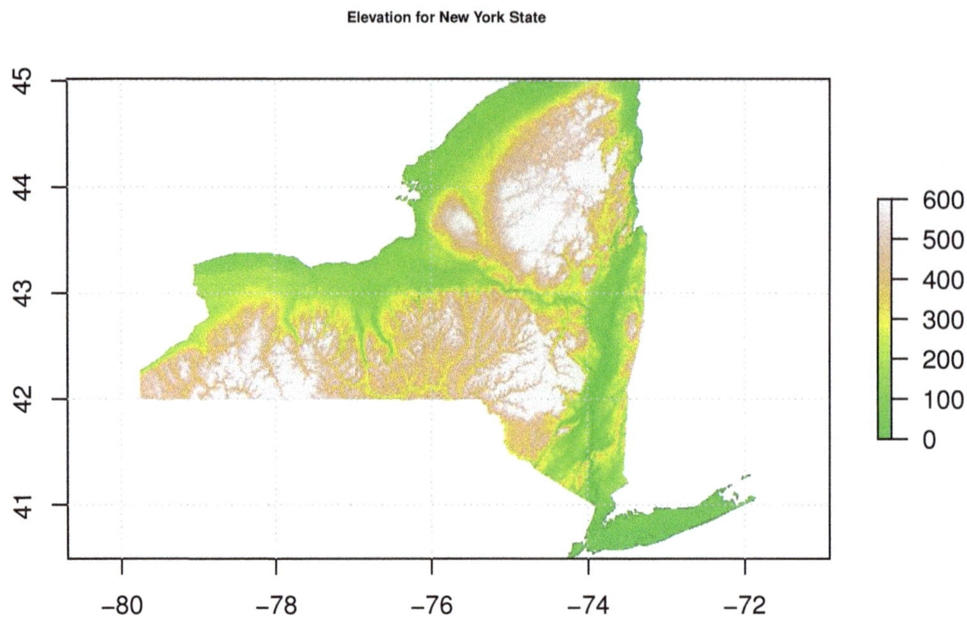

Figure 3.9: Elevation for New York State clipped to the 0 to 600 m range.

White areas represent elevations higher than 600 m because the map display was clipped to the 0 to 600 m range.

We now move from a state-scale view to a smaller coastal region where terrain detail becomes more important.

DEM for Long Island

Zooming into Long Island requires downloading elevation data with a higher zoom level (`z`) to increase detail. Here we zoom in on Long Island by using a smaller study-area boundary instead of the entire New York State boundary. Long Island is generally low-lying, making it an ideal study area for coastal terrain and flooding analysis (Figure 3.10).

Coastal areas are particularly important for understanding flood risk and coastal processes, while slightly higher inland elevations may imply lower risk from coastal flooding but could still be affected by other hazards such as poor drainage, inland flooding, or sinkholes.

```
# Get elevation data for Long Island
# Use the original geographic-coordinate li boundary for downloading
# Many online elevation services internally expect geographic coordinates
li_dem <- elevatr::get_elev_raster(
  locations = li,
```

```
  z = 9,
  clip = "locations"
)
# Plot DEM with elevation constraints
plot(
  clamp(li_dem, lower = 0, upper = 100),
  main = "Elevation above Sea Level for Long Island Region",
  cex.main = 0.6,
  col = terrain.colors(30)
)
grid()
```

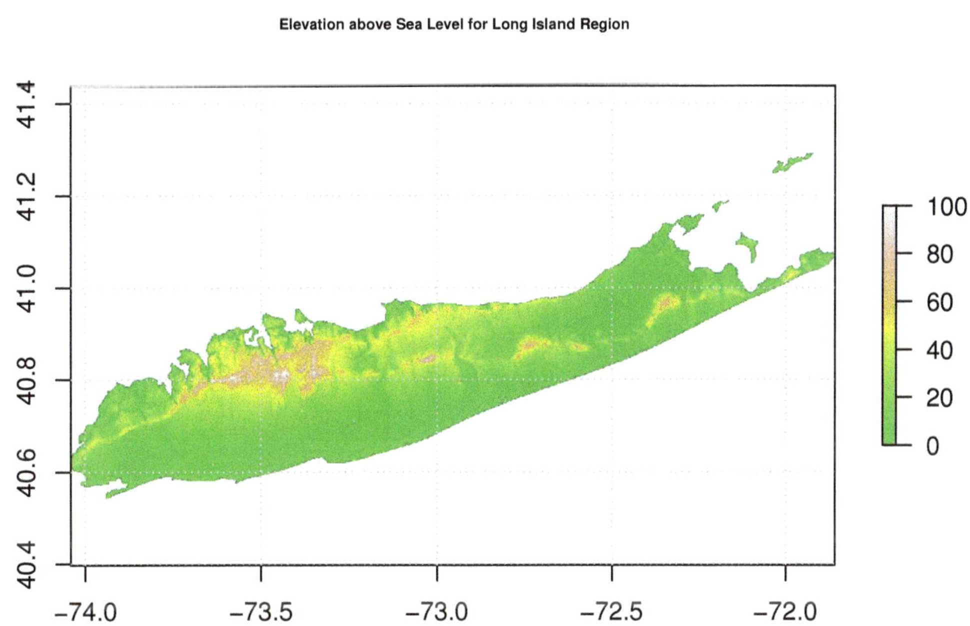

Figure 3.10: Elevation for Long Island above sea level.

> **i Why this Long Island example matters**
>
> This Long Island DEM becomes the starting point for the flood-simulation case study in Chapter 6.

Bathymetry of New York Bight

By changing the range of elevation values we display, we can emphasize different parts of the terrain. So far, we focused on elevations above sea level. We now shift below sea level to examine bathymetry (the underwater version of elevation, describing the shape and depth of the seafloor) in the New York Bight, the curved stretch of coastal ocean off New York and New Jersey. Elevations below 0 m correspond to areas below sea level, while low elevations (0 to 40 m) are often important in coastal and flooding studies. Figure 3.11 emphasizes areas at or below sea level, which are especially relevant for coastal and flood-related studies. We now zoom in on the Long Island region and focus on bathymetry to observe Long Island Sound (Figure 3.12).

```
# Get elevations around New York Bight
ny_bight <- elevatr::get_elev_raster(
  # retrieve elevation data around Long Island region
  locations = li,
  z = 7 # resolution from 1-14, higher is more detailed
)
# Plot the DEM with elevation constraints:
# Focus on low elevations at or below sea level (bathymetry)
# Everything above 20 meters is removed
plot(
  clamp(ny_bight, lower = -80, upper = 20),
  xlim = c(-74.5, -71.5),      # longitude range
  ylim = c(39.1, 41.3),        # latitude range
  main = "Areas at or Below Sea Level",
  cex.main = 0.6,
  col = terrain.colors(30)
)
grid()
```

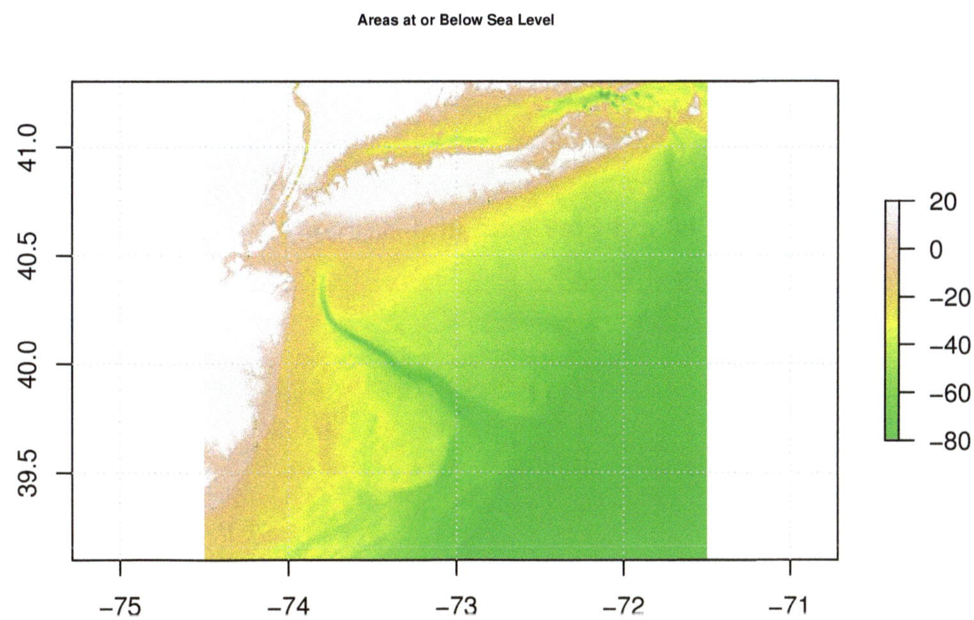

Figure 3.11: Areas at or below sea level in the New York Bight region.

The resulting bathymetric DEM reveals the underwater topography of the New York Bight. We now repeat the process, but this time we increase the resolution and constrain our spatial extent around Long Island Sound.

Bathymetry of Long Island Sound

```
# Get elevation data around Long Island Sound
ny_bath <- elevatr::get_elev_raster(
  locations = li,
  z = 10
)
# Plot a bathymetric map of Long Island Sound
# Focus on low elevations at or below sea level (bathymetry)
# Everything above 20 meters is removed
plot(
  clamp(ny_bath, lower = -80, upper = 20),
  xlim = c(-74.5, -71.5),      # longitude range
  ylim = c(40.4, 41.3),        # latitude range
  main = "Long Island Elevation and Bathymetry",
```

```
  col = terrain.colors(30)
)
grid()
```

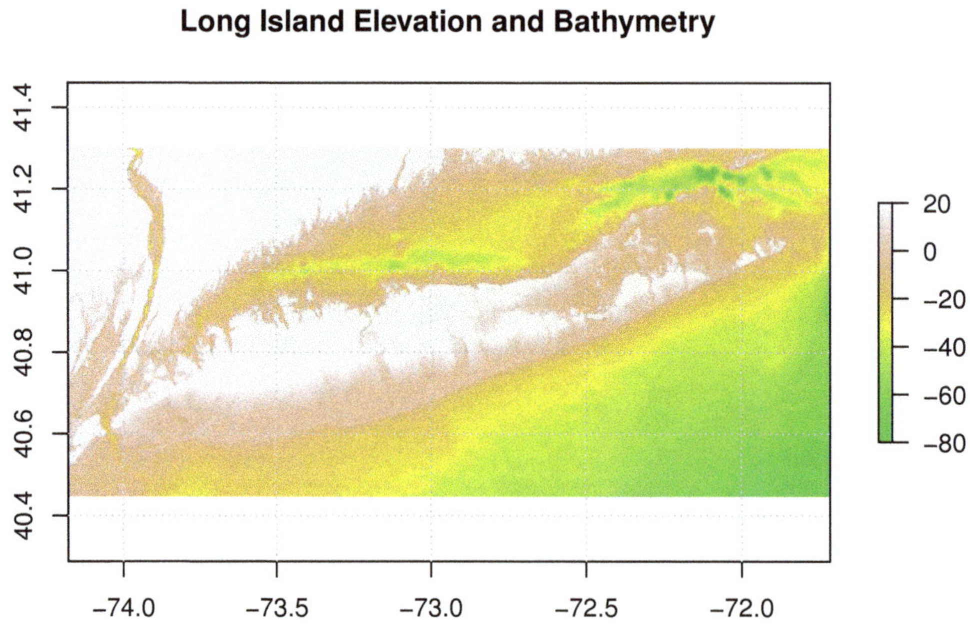

Figure 3.12: Low-elevation and below-sea-level areas near Long Island.

The continental shelf and shallow coastal bathymetry become clearer at this higher resolution (Figure 3.12). The smooth transition from shallow (lighter colors) to deeper areas (darker colors) reveals the rapid drop-off into deeper ocean waters.

> **i Why elevation matters**
>
> In Chapter 6, we return to this DEM to identify flood-prone areas in Long Island, NY, connecting terrain analysis with climate resilience and sustainability planning.

Histogram of elevation values

The raster map shows where the land is higher or lower, while the histogram (Figure 3.13) shows how those elevations are distributed across the state. Most of the state has low to moderate elevation, with a long tail representing mountainous regions.

```
# Using New York State DEM we will
# extract elevation values and clean data for plotting

# Extract elevation values from the raster
ny_vals <- values(ny_dem)
# Remove NA values for plotting
ny_vals <- ny_vals[is.finite(ny_vals)]
# Create a data frame for histogram
ny_df <- data.frame(elevation_m = ny_vals)
# Plot a histogram of elevation values
hist(
  ny_df$elevation_m,
  breaks = 40,
  col = "lightblue",
  border = "white",
  main = "Elevation Distribution for New York State",
  xlab = "Elevation (m)",
  ylab = "Cell count"
)
```

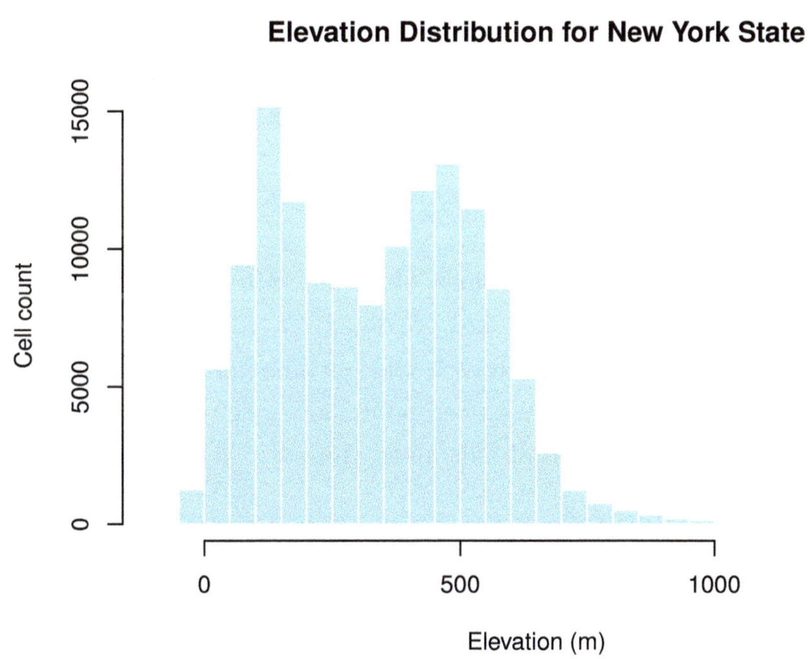

Figure 3.13: Distribution of elevation values across New York State.

> **i** Two ways to read a raster
>
> The map shows where elevations occur, while the histogram shows how those values are distributed. Together, they give a clearer and more complete picture than either one alone.

This combination of spatial pattern and value distribution is a key idea that appears throughout geoscience analysis.

3.3.11 From a Spatial Object to a Topographic Profile

Sometimes we need to convert spatial data into a regular plot. In the next example, we will extract elevation values from a DEM along a line across Long Island and convert those values into a chart called a topographic profile.

An `sf` object can behave like a plain table for attribute work, but for a standard chart we often remove the location/shape part first with `st_drop_geometry()`, keeping only the data columns we need. In contrast, `extract()` uses geometry (location) to obtain new values from a raster, such as elevation values sampled along a line.

3.3.12 Topographic Profile from a DEM

A DEM can be sampled along a line to create a simple topographic profile. Here we define two points, connect them with a line, extract elevation values from the DEM, and then plot elevation against distance as a regular chart.

Once data are projected, all coordinates must use the same units (meters), not longitude and latitude. Because the line is defined in a projected coordinate system, we can measure distance in kilometers and create a meaningful elevation profile.

The red line on the map below defines the path where elevation values will be sampled from the DEM. Again, we will use already loaded libraries and previously saved variables.

What we are doing here with the code: We will draw a straight line across Long Island from west to east, sample the elevation value at 100 evenly spaced points along that line, and then plot those values as a chart showing how the land rises and falls along the path. This converts spatial raster data into a simple x-y plot.

```r
# Create a line using two geographic coordinate points
# (longitude, latitude)
line_wgs84 <- sf::st_sfc( # convert to sf object
  st_linestring(
    rbind(
      # Each c() stores one longitude/latitude location
      c(-74.05, 40.65),  # western point
```

```
    c(-71.80, 41.05)   # eastern point
  )
 ),
 crs = 4326 # geographic coordinates
)

# Project line to UTM 18N
line_utm <- sf::st_transform(line_wgs84, 32618)

# Reuse the DEM created earlier in the chapter
# Convert the DEM into a terra raster
li_dem <- rast(li_dem)
# Project the DEM to the same CRS as the line (UTM 18N)
li_dem_utm <- terra::project(li_dem, "EPSG:32618")

# Plot the DEM
plot(
  clamp(li_dem_utm, upper = 200, lower = 0),
  xlim = c(580000, 760000),    # west → east (meters)
  ylim = c(4480000, 4570000), # south → north (meters)
  col = terrain.colors(30),
  main = "Topographic Profile Transect Across Long Island"
)
# Add the transect line
plot(
  vect(line_utm),
  add = TRUE,
  col = "red",
  lwd = 3
)
```

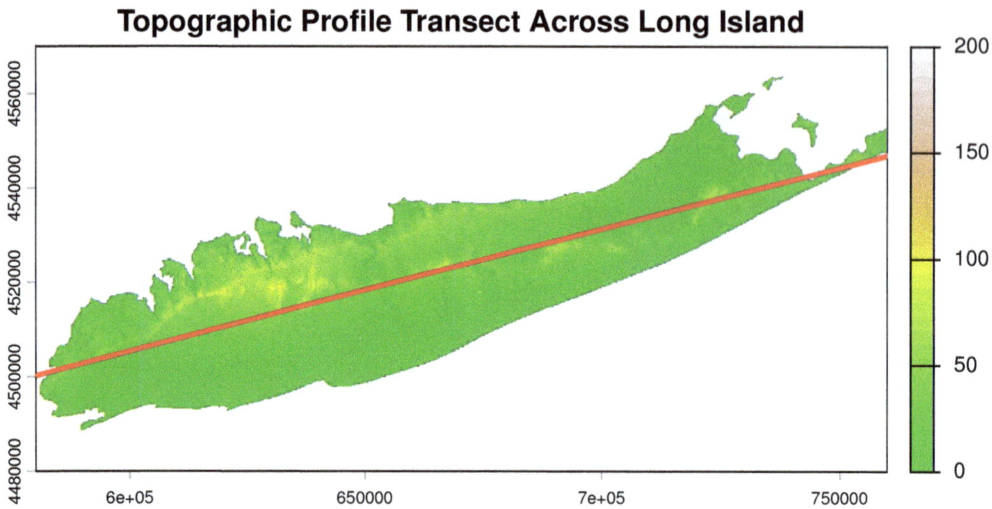

Figure 3.14: West to east transect across Long Island used to extract a topographic profile.

Because the line is in a projected coordinate system, we can measure its length and express the profile in real distance (kilometers) instead of relative units. The sampled elevation values are then plotted as a topographic profile in Figure 3.15.

Some elevation functions return raster data in a format that must first be converted into a **terra** raster object before we can use **terra** analysis tools.

```
# Sample 100 points along the projected line
line_pts <- st_line_sample(line_utm, n = 100)
# Treat each dot as a separate point we can sample
line_pts <- st_cast(line_pts, "POINT")
# Create an sf object with the sampled points
line_sf <- st_sf(id = seq_along(line_pts), geometry = line_pts)

# Convert DEM
# to terra raster format
ny_dem <- rast(ny_dem)
# Project the DEM to the same CRS as the line
dem_utm <- terra::project(ny_dem, "EPSG:32618")

# Extract elevation values at the sampled points
elev <- terra::extract(dem_utm, vect(line_sf))
```

```r
# Build a data frame
# Convert meters to km
total_length_km <- as.numeric(st_length(line_utm)) / 1000
# Create topographic profile
profile_df <- data.frame(
  dist_km = seq(0, total_length_km, length.out = nrow(elev)),
  elevation = elev[, 2]
)

# Plot profile
plot(
  profile_df$dist_km,
  profile_df$elevation,
  type = "l",
  xlab = "Distance (km)",
  ylab = "Elevation (m)",
  main = "Topographic Profile Across Long Island"
)
```

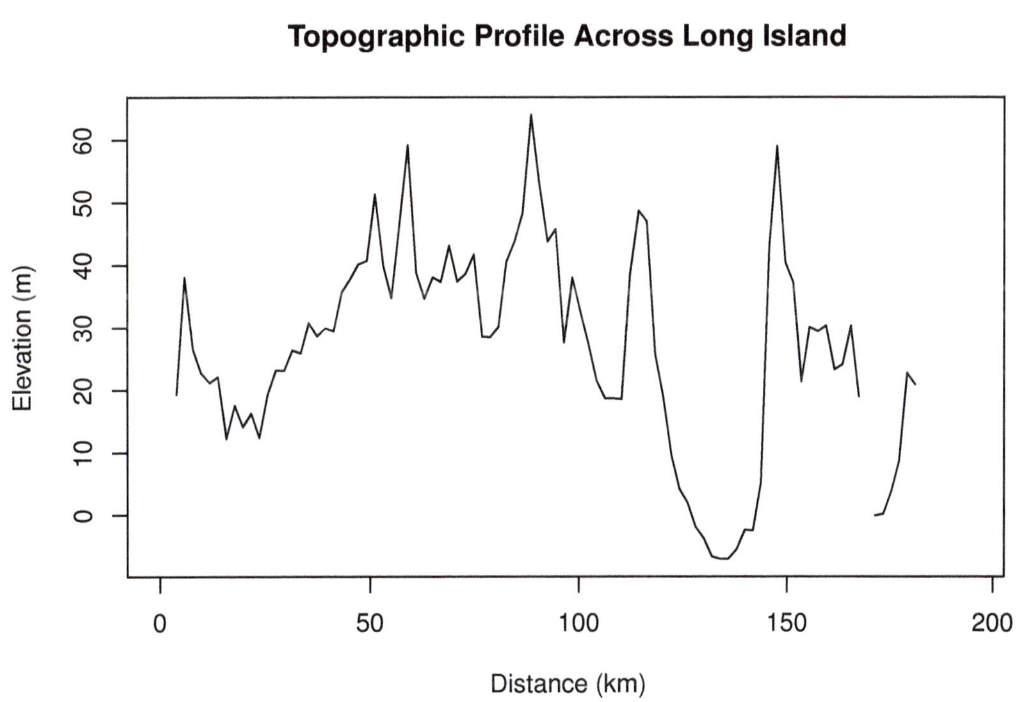

Figure 3.15: Topographic profile extracted along the transect across Long Island.

Here we are no longer drawing a map; we are using the attribute table inside the spatial object to make a regular plot.

> **i From geometry to a table**
>
> The geometry is used to extract elevation values from the raster. After extraction, we work with a regular table (data frame) of values for plotting. There is no geometry inside the data frame `profile_df`.

3.3.13 Review Simple Spatial Operations

Once a spatial object is loaded, we can do simple operations such as:

- cropping to a boundary
- masking data outside a study area
- selecting features by name or attribute
- summarizing values within a region
- combining (union) several polygons into one
- extracting spatial data using features from a DEM raster

3.3.14 Clip a Raster to a Study Area

As a brief review, we can crop the raster to a boundary to keep only the study area. We use the state boundary to keep only the area we care about.

> **i Working with `sf` and `terra`**
>
> Spatial data can be stored in different formats depending on the package.
>
> - `sf` is mainly used for vector data (points, lines, polygons)
> - `terra` is used for raster data and also has its own vector format (`SpatVector`)
>
> Both packages can represent vector data, but they use different object types. Because of this, some functions require conversion between formats.
> For example:
>
> - `ny_state` → an `sf` object
> - `vect(ny_state)` → the same geometry converted to a `terra` vector
>
> We convert to `terra` when working with raster functions such as clipping or extracting values from a DEM.
> Common conversions:

- `vect()` → `sf` → `terra`
- `st_as_sf()` → `terra` → `sf`

It's like using different file formats: sometimes you need to convert data so another tool can read it.

```
# Convert the New York state boundary
# into a vector format that raster tools can use
ny_state_vect <- vect(ny_state)

# Crop the raster so the analysis focuses on the New York study area
ny_dem_clip <- crop(ny_dem, ny_state_vect)

plot(
  ny_dem_clip,
  main = "New York State DEM Clipped to Boundary",
  cex.main = 0.6,
  col = terrain.colors(20)
)
grid()
```

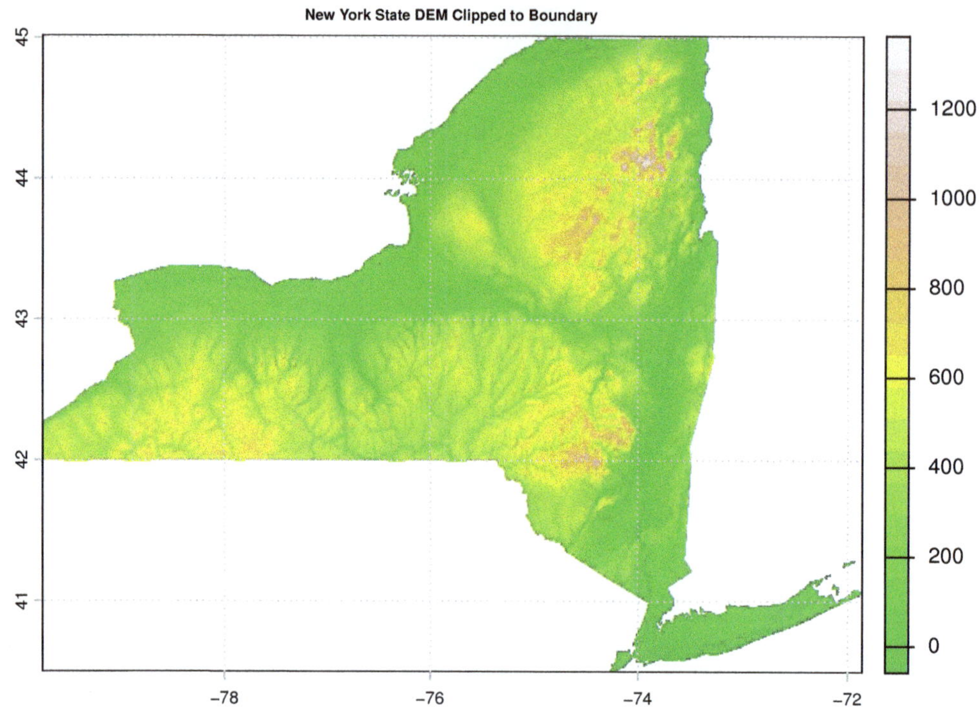

Figure 3.16: Elevation raster clipped to the New York State boundary.

This step restricts the raster to the New York study area so the analysis focuses only on the relevant region (Figure 3.16). The vector `ny_state_vect` defines the study area for the clipping.

- Raster = a big sheet of cookie dough
- Vector = a cookie cutter (shape)

A vector boundary can be used to keep only part of a raster.

3.3.15 Writing Clear Spatial Code

The most important habit in spatial work is to write code that is easy to read later.

3.3.16 Good practice checklist

- Use clear object names.
- Add comments to explain why each step is done.
- Keep the code reproducible.
- Separate data loading, transformation, analysis, and plotting.
- Use the same coordinate system before comparing layers.

> Think twice, code once
>
> Code is easier to understand when it is written in small, clear steps. Good object names help explain what the data represents.

Up to this point, we have focused on spatial tools individually. In the next section, we combine these ideas into a complete terrain-analysis workflow using Long Island as a study area.

3.4 Part C: Advanced Applications

Part C is a more advanced example. Focus on the idea, not every line of code. It moves from basic GIS ideas to a practical coastal application. We will build a Long Island study area, download elevation data, and create a hillshade to better visualize the terrain. The goal is to prepare elevation data that can later be used for more advanced analysis.

3.4.1 Building a Study Area and Creating a DEM

We will use `sf` for vector data manipulation, `terra` for raster analysis and hillshade creation, and `elevatr` for downloading elevation data. The workflow includes:

Step 1: Define study area

Long Island is represented here by the four counties commonly used in GIS analyses of Long Island: Kings, Queens, Nassau, and Suffolk. First, we read the county boundaries. Then we clean the shapes, move them into the right coordinate system, and combine them into one Long Island outline.

```
# Load libraries
library(elevatr) # for downloading elevation data
library(tigris) # for downloading boundary data
library(terra)    # for working with raster data
library(sf) # for working with spatial vector data

# Download U.S. counties for New York State
# at a resolution (500k)
# Keep in mind that this layer starts in geographic coordinates
# (longitude/latitude)
li <- tigris::counties(state = "NY", cb = TRUE, resolution = "500k", year = 2024)
# Select Long Island counties
li <- li[li$NAME %in% c("Kings", "Queens", "Nassau", "Suffolk"), ]

# Transform to UTM first for accurate distance measurements
li_utm <- st_transform(li, 32618)

# Grow the land outline outward by 5 km to make a slightly larger area
buffer_m <- 5000 # 5 km, written in meters
study_area <- st_buffer(li_utm, dist = buffer_m)

# Subtract the land from that larger area, leaving only the surrounding
```

```
# water ring
water_utm <- st_difference(study_area, li_utm)
water_utm <- st_collection_extract(water_utm, "POLYGON")
water_utm <- st_make_valid(water_utm)

# Plot water first (background)
plot(
  st_geometry(water_utm),
  col = "lightblue",
  border = NA,
  axes = TRUE,
  main = "Long Island Land + Water Polygon"
)
# Add land layer
plot(
  st_geometry(li_utm),
  col = "tan",
  border = "black",
  lwd = 1,
  add = TRUE
)
grid()
```

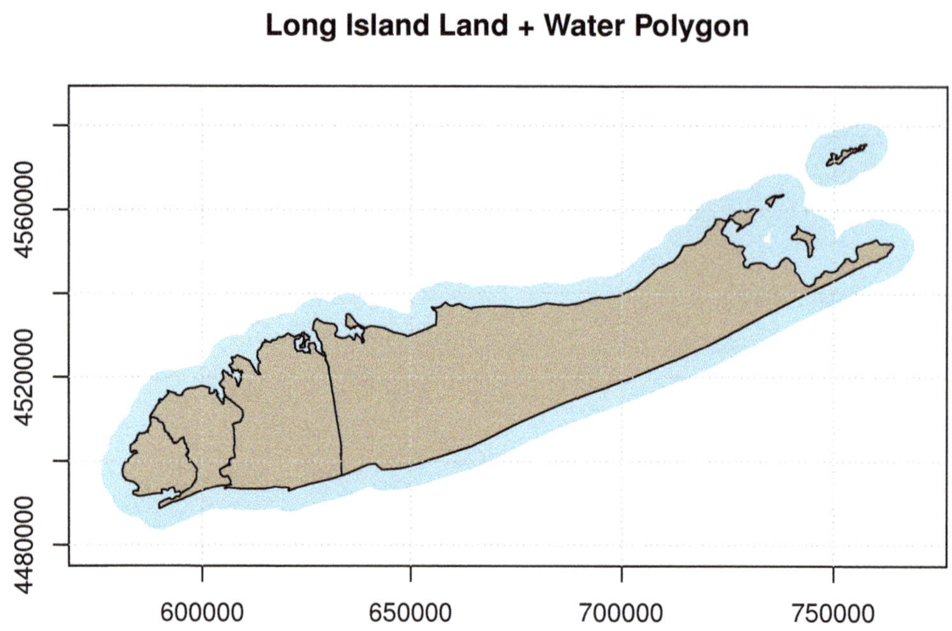

Figure 3.17: Long Island study area showing land and surrounding water buffer.

The number `32618` is an EPSG code, which is like an ID number for a coordinate system.

Interpretation: Figure 3.17 provides a clean outline of Long Island and the surrounding water buffer. It consists of multiple polygons (counties), and it is the geographic frame we will use for the DEM analysis. The axes now show projected coordinates in meters instead of longitude and latitude degrees. This is because the data have been reprojected into a coordinate system designed for measurement.

First, we need to dissolve all polygons into one shape to create a single outline of Long Island. This is done with `st_union()`, which combines all the county polygons into one. The resulting shape is a single polygon that represents the entire land area of Long Island, which is important for later steps when we want to clip the DEM to this area of interest or to simulate flooding in this specific region.

When several polygons are merged into one shape, R no longer knows which original attributes should be kept (for example, which county name or population values should remain associated with the new geometry). To avoid mixing or duplicating information incorrectly, the attribute table is temporarily removed.

We then use `st_sf()` to rebuild the object as a spatial object again. This new object contains the geometry of the dissolved shape together with a new attribute column (in this case, just a name for the whole area). This allows us to keep both the geometry (the shape of Long Island) and the attribute information organized together.

⚠ Dissolving polygons

Dissolving polygons combines multiple shapes into one. This is useful when you want to treat a group of features as a single unit, such as combining all counties into one Long Island shape. The key thing to remember is that the **sf** property is lost after union, so we need to restore it with `st_sf()`.

ℹ Same projection

When we project multiple layers (e.g. land and water), make sure they are all in the same coordinate system. This ensures that they will align correctly when plotted together or used in analysis. **st_transform** will reproject the data to a new coordinate system (UTM), which is necessary for accurate spatial analysis and measurement.
A common workflow is:

- use geographic coordinates (longitude and latitude) to retrieve or locate data,
- then transform everything into a projected system (UTM) for analysis and measurement.

```
# Dissolve into one Long Island polygon
li_utm <- st_union(li_utm) # Combine the four counties into one shape

# st_union removes some sf object structure, so we restore it with st_sf
li_utm <- st_sf(name = "li_utm", geometry = li_utm)

# let's check the object
head(li_utm)
```

```
Simple feature collection with 1 feature and 1 field
Geometry type: MULTIPOLYGON
Dimension:     XY
Bounding box:  xmin: 581047.4 ymin: 4488651 xmax: 764108.9 ymax: 4575779
Projected CRS: WGS 84 / UTM zone 18N
    name                       geometry
1 li_utm MULTIPOLYGON (((581114.5 44...
```

Notice that the object contains both attribute columns and a geometry column. This is what makes it a spatial object instead of a regular data frame.

```
# Plot the dissolved Long Island shape
# Plot water first (background)
plot(
  st_geometry(water_utm),
  col = "lightblue",
  border = NA,
  axes = TRUE,
  main = "Long Island Land (dissolved counties) + Water Polygon"
)
# Add land on top
plot(
  st_geometry(li_utm),
  col = "tan",
  border = "black",
  lwd = 1,
  add = TRUE
)
grid()
```

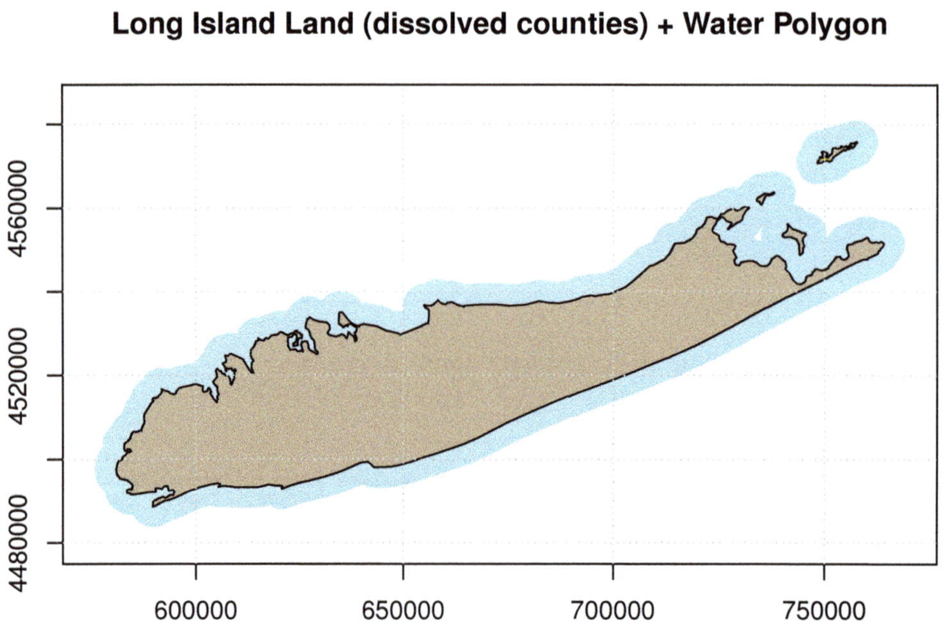

Figure 3.18: Long Island study area with county boundaries dissolved into one polygon.

Now that we have dissolved all counties into one polygon, and transformed all layers (the layers `water_utm` and `li_utm`) to UTM (meters), we can use this polygon as our final study boundary. This is the area we will use to clip the DEM and to analyze coastal processes in later chapters. The map (Figure 3.18) shows the new Long Island shape after dissolving the county boundaries, with the surrounding water buffer still visible. The axes now show meters instead of degrees, indicating that the data have been reprojected into a coordinate system suitable for spatial analysis.

Step 2: Download a digital elevation model (DEM)

We will download a digital elevation model of Long Island using the study-area boundary, limiting the download to that area with the `clip = "locations"` parameter. Even if the boundary is projected in UTM, many online elevation services internally use geographic coordinates (WGS84 longitude and latitude) and often return the DEM in WGS84.

> **i Location or measurement?**
>
> Geographic coordinates (degrees) are good for location, while projected coordinates (meters) are better for measurement. Many external elevation services provide DEMs in WGS84 longitude and latitude by default. For analysis, we usually transform the DEM into a projected coordinate system such as UTM so distances and areas can be measured accurately. The `project()` function from the `terra` package is used to reproject the DEM to UTM zone 18N (EPSG:32618), which is suitable for the Long Island region.

```
# We will reuse the DEM already created
# earlier in the chapter (li_dem_utm)

# Plot DEM alone
plot(
  clamp(li_dem_utm, lower = 0, upper = 100),
  col = terrain.colors(30),
  main = "Digital Elevation Model (DEM)",
  cex.main = 0.8
)
grid()
```

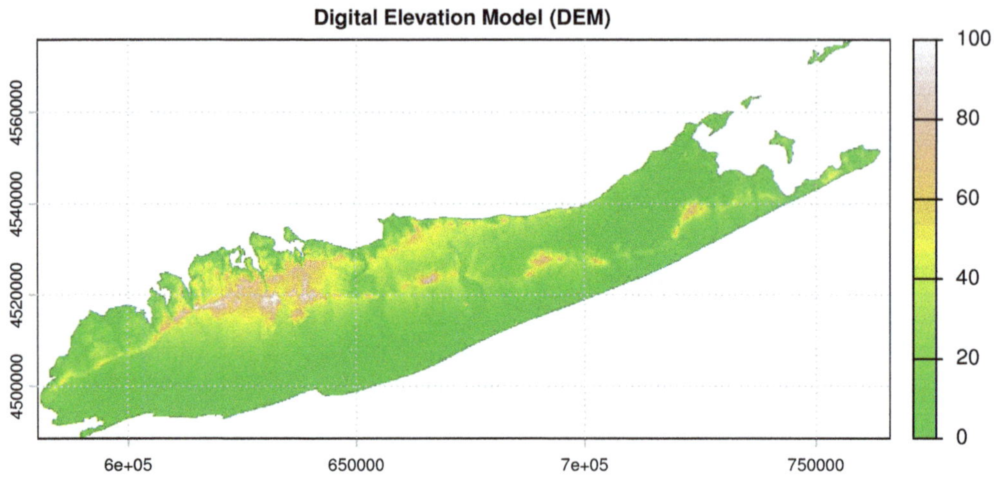

Figure 3.19: Digital elevation model of Long Island showing subtle coastal elevation variation.

Subtle elevation variation across Long Island is critical for understanding coastal processes (Figure 3.19). Keep in mind that elevation data are often returned in geographic coordinates (WGS84 longitude and latitude). After downloading, we convert the data to a projected system (UTM) for analysis.

> **i** What the DEM shows
>
> Low elevation areas are important in many geoscience applications. A DEM does not show water directly, but it helps us understand terrain and elevation patterns.

Interpretation: The DEM shows where Long Island is higher and lower, helping us understand how terrain influences flooding, drainage, and coastal processes. Even small elevation differences can strongly influence coastal flooding and drainage patterns. Lower coastal areas appear as greener colors. Inland regions gradually transition to yellow and reddish colors at higher elevations. A few isolated white regions represent the highest terrain. This gradual rise away from the shoreline is typical of many coastal landscapes and helps explain why some areas are more vulnerable to coastal flooding, storm surge, and poor drainage than others.

Changing DEM colors: Sometimes we need to change DEM colors to better visualize the terrain. The `clamp()` function is used to limit the range of elevation values displayed, which can help highlight specific features. For example, if we are interested in coastal areas, we might focus on elevations between 0 and 100 meters, which are most relevant for flood risk and coastal processes. By adjusting the color scheme and value range, we can make important features more visible and easier to interpret. Below, we will change the color scheme to start with blue colors for low elevations and then transition to green, yellow, and red for higher elevations. This color scheme can help us quickly identify areas of interest based on elevation.

```
# Plot DEM alone with
# a different color scheme and value range
plot(
  clamp(li_dem_utm, lower = 0, upper = 100),
  col = topo.colors(30),
  main = "Digital Elevation Model (DEM)",
  cex.main = 0.8
)
grid()
```

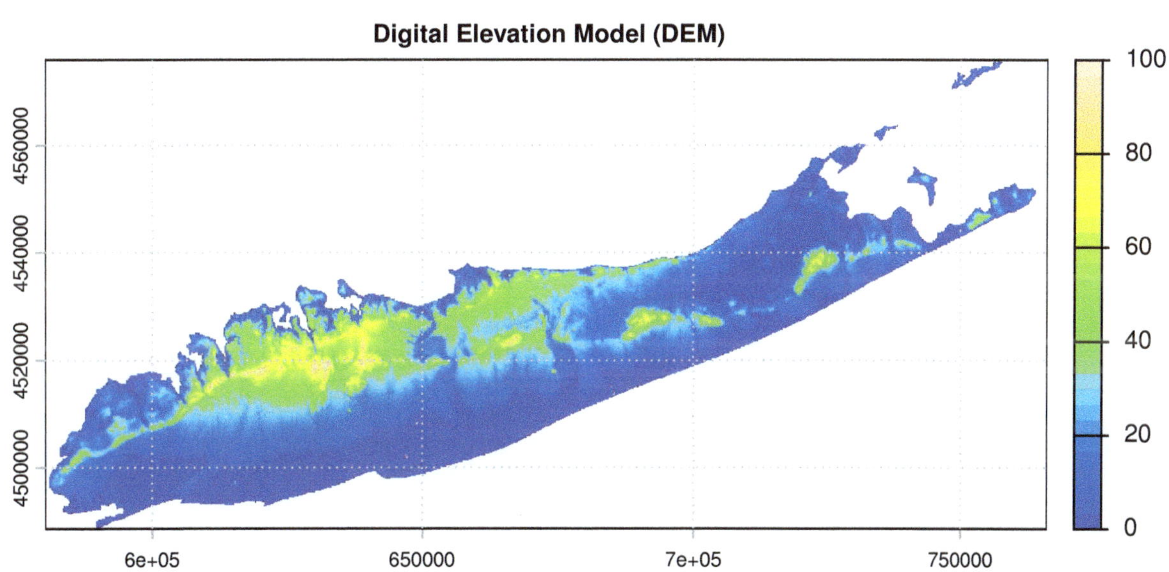

Figure 3.20: Long Island digital elevation model shown with a topographic color scale (0 to 100 m).

Step 3: Visualize terrain with coastline context

We draw the elevation first, then add the coastline and land boundaries on top. This layering approach helps us interpret the terrain in relation to the coast.

The coastline overlay adds spatial context to the elevation data, helping us interpret how terrain interacts with coastal boundaries and where low-lying areas occur near the shoreline.

```
# Add water (semi-transparent)
# always projected first to ensure it aligns with the DEM
plot(
  st_geometry(water_utm),
  col = adjustcolor("lightblue", alpha.f = 0.5),
  border = NA
)

# Plot DEM
plot(
  clamp(li_dem_utm, lower = 0, upper = 100),
  main = "Long Island DEM with Coastline",
  axes = TRUE,
  col = terrain.colors(30),
  add = TRUE
)

# Add land boundary
plot(
  st_geometry(li_utm),
  border = "black",
  lwd = 2,
  add = TRUE
)
grid() # Add a grid for better spatial reference
```

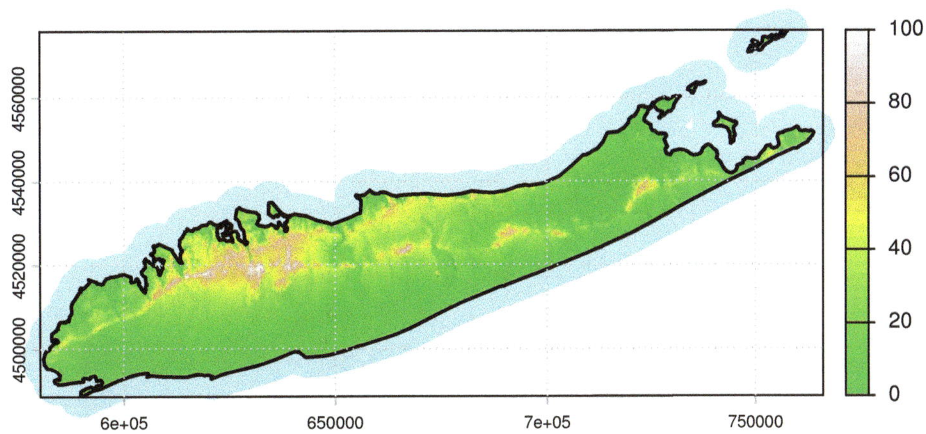

Figure 3.21: Long Island DEM with coastline overlay for spatial context.

Interpretation: Figure 3.21 combines elevation with coastline context, making it easier to interpret how terrain interacts with the coastal boundary. The coastline helps us see whether the raster is aligned properly and where the lowest areas are located.

Step 4: Create a hillshade to enhance terrain visualization

A hillshade is a shaded relief map that simulates how light would interact with the terrain. It creates shadows and highlights that make hills and valleys easier to see. This is especially helpful in flat coastal areas where the DEM alone can look visually plain. The hillshade is computed from how steep the land is (slope) and which way each point faces (aspect), then brightened to make the differences easier to see.

Vertical exaggeration is applied to the DEM to make subtle elevation differences more visible, especially in flat coastal areas.

The hillshade is then masked to match the study area while preserving the same projected coordinate system as the DEM. The resulting hillshade map (Figure 3.22) brings out terrain relief and surface structure more clearly, making it easier to identify features such as ridges, valleys, and coastal landforms.

⚠ Raster functions require raster objects

Be aware that if the DEM is not converted to raster (using the **terra** package with the function **rast()**) then terra functions like **shade** will not be able to produce the hillshade. Always remember that **sf** and **terra** use different object types for vector and raster operations.

```
# Exaggerate elevation to make
# relief easier to see in flat coastal terrain
dem_ex <- li_dem_utm * 5

# Compute terrain properties
slope  <- terra::terrain(dem_ex, "slope", unit = "radians")
aspect <- terra::terrain(dem_ex, "aspect", unit = "radians")

# Better lighting for perception
hillshade <- terra::shade(
  slope, aspect,
  angle = 40,
  direction = 315
)

# Contrast stretch: find the darkest and brightest 2% so a few extreme
# values do not dominate the picture, then spread the rest across the full
# range
hs_vals <- values(hillshade)
q <- quantile(hs_vals, probs = c(0.02, 0.98), na.rm = TRUE)

hillshade_stretch <- clamp(hillshade, q[1], q[2])
hillshade_stretch <- (hillshade_stretch - q[1]) / (q[2] - q[1])

# Project the hillshade and boundary
# to the same CRS (UTM) before masking
hillshade_utm <- terra::project(hillshade_stretch,
                                "EPSG:32618",
                                method = "bilinear")

# Convert the sf polygon into a terra vector object
# because terra raster functions expect terra vector format
li_v <- vect(li_utm)
# crop hillshade to Long Island area so the map is tighter
hillshade_utm <- mask(hillshade_utm, li_v)
```

```
plot(
  hillshade_utm,
  legend = FALSE,
  axes = FALSE,
  box = FALSE,
  col = gray(seq(0, 1, length.out = 100))
)
grid()
```

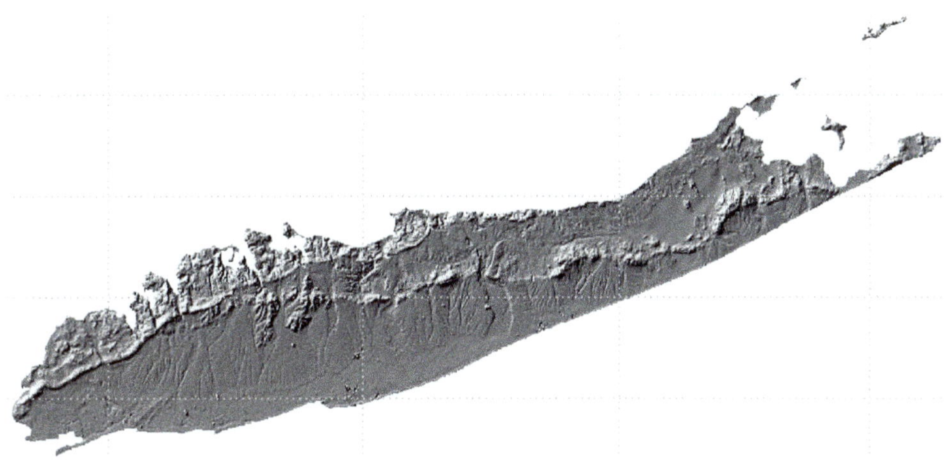

Figure 3.22: Hillshade projected and masked to match the study area.

The `vect()` function converts an `sf` spatial object into a `terra` vector object (`SpatVector`) so it can interact correctly with raster functions from the `terra` package.

Interpretation: The hillshade in Figure 3.22 enhances subtle terrain variations that are difficult to recognize in the DEM alone. The raised ridges visible across Long Island are glacial landforms created during the last Ice Age. As glaciers moved, they pushed and left behind sediment, creating these elevated areas. These features explain why some areas are higher than others and help us understand how water may flow or collect across the island.

> **i Hillshade makes terrain easier to interpret**
>
> Hillshade adds shadows, not new data.

The next example is more advanced.

Do not focus on memorizing every function. Focus on the main idea:

- create terrain shading,
- combine it with elevation,
- and make landforms easier to visualize.

Step 5: Add a DEM on the hillshade and a scale

Combining a hillshade with a DEM creates a more three-dimensional visualization of terrain. Before combining the hillshade and the Long Island DEM, we project both layers to the same coordinate system so they align correctly. Finally, a vertical exaggeration is applied to the DEM (by multiplying the DEM values) to make the relief more visible, especially in flat coastal areas where elevation differences are subtle. The resulting map (Figure 3.23) shows the hillshade with the DEM overlaid, providing a clearer visualization of terrain relief and surface structure across Long Island.

```
# Exaggerate elevation to make relief easier to see in flat coastal terrain
dem_ex <- li_dem_utm * 5
```

This does not change the real elevation values permanently. It temporarily exaggerates the terrain so small elevation differences become easier to see visually.

```
# Compute terrain properties
slope  <- terra::terrain(dem_ex,
                         "slope",
                         unit = "radians")
aspect <- terra::terrain(dem_ex,
                         "aspect",
                         unit = "radians")

# Better lighting for perception
hillshade <- terra::shade(
  slope, aspect,
  angle = 40,
  direction = 315
)
```

```
# Contrast stretch
hs_vals <- values(hillshade)
q <- quantile(hs_vals, probs = c(0.02, 0.98),
              na.rm = TRUE)

hillshade_stretch <- clamp(hillshade, q[1], q[2])
hillshade_stretch <- (hillshade_stretch - q[1]) / (q[2] - q[1])
# Confirm same projection (CRS)
# Reprojecting dem_ex or hillshade_stretch to the same CRS (UTM)
# is not necessary here
# because dem_ex already originated from a UTM DEM
hillshade_stretch_utm <- hillshade_stretch
dem_ex_utm <- dem_ex
```

Remember to check that all your layers use the same projection; otherwise your map will not be accurate.

```
st_crs(dem_ex_utm)
st_crs(hillshade_stretch_utm)
```

Geographic coordinate systems often reference WGS84 (`World Geodetic System 1984`), while projected coordinate systems may use UTM zones such as `UTM Zone 18N` identified by EPSG codes like `32618`.

```
# Plot all in one map
plot(hillshade_stretch_utm,
     col = gray(0:100/100),
     legend = FALSE,
     main = "Hillshade Visualization of Long Island DEM")
plot(clamp(dem_ex_utm, lower = 0, upper = 200),
     alpha = 0.7,
     add = TRUE) # Add the DEM on top
grid()
# Add legend for scale
segments(736000, 4500000, 756000, 4500000, lwd = 2)
text(747000, 4504000, "20 km", cex = 0.8)
# North arrow
arrows(750000, 4510000, 750000, 4520000, lwd = 2, length = 0.1)
text(750000, 4522000, "N", cex = 1)
```

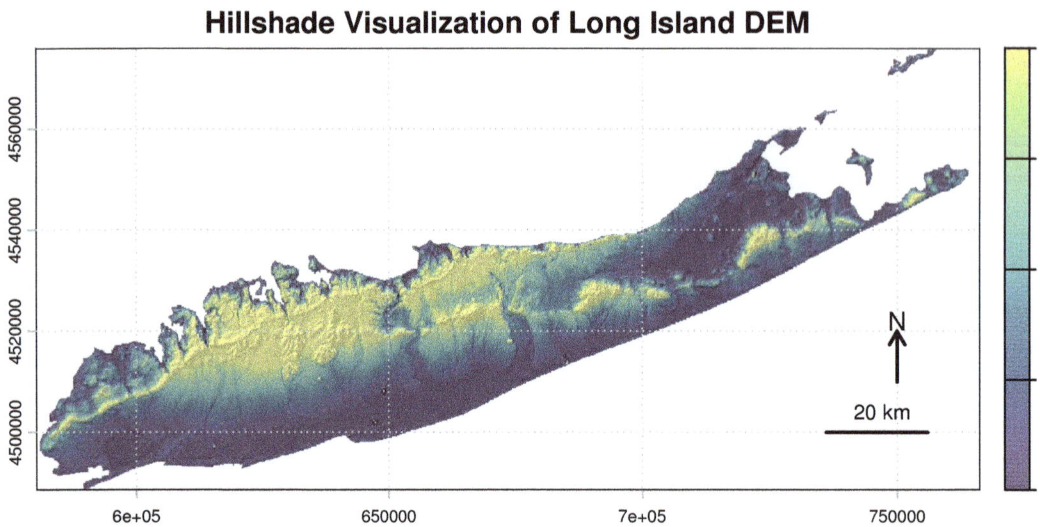

Figure 3.23: Hillshade visualization enhancing terrain relief and surface structure.

Because the map uses a projected coordinate system in meters, distances can be measured directly and displayed with a scale bar. If the map were still using longitude and latitude, the scale would vary across the map because degrees are not constant distance units. A simple way to think about this is that longitude and latitude divide the curved Earth into sections, similar to orange peels wrapped around a sphere. Those sections help us locate places, but the distances inside them are not constant everywhere on Earth. Projected coordinate systems flatten the curved surface into a map where distances can be measured more consistently.

Adding a scale is easy with a UTM projection by subtracting two coordinates, and it allows us to understand real-world distances on the map. The scale bar indicates that the distance between the two points (`736000` and `756000`) is 20000 meters, that is 20 km, which helps us interpret the size of features and the extent of the study area. Accurate scale information matters for understanding spatial relationships and future analysis related to coastal processes or flood risk.

Interpretation: Projecting the hillshade (Figure 3.22) ensures it aligns correctly with the DEM within the study area. If the layers do not share the same coordinate system, the map can look wrong even if each layer is correct on its own.

> ⚠ A map can look right and still be wrong
>
> Always check projection, extent, and units before combining layers. Visual alignment alone is not enough.

ℹ What we have built

- A defined study area (Long Island)
- A processed DEM (elevation surface)
- A visualization of terrain structure

In Chapter 6, we will use this exact dataset to:

- identify low-lying areas
- estimate flooded area
- interpret coastal risk

ℹ Preparing for flood analysis

We are not mapping floods yet; we are preparing the terrain.

3.4.2 Looking Ahead

The DEM we created here will be used again later in the book. In Chapter 6, we will return to elevation data and use it in a more advanced coastal flooding case study.

GIS methods used for rivers, elevation, and boundaries also apply to geological datasets such as faults, rock units, and tectonic structures. Once you understand raster and vector workflows, geological layers become another type of spatial data that can be analyzed in the same way.

3.5 Common Mistakes to Avoid

- **Mixing coordinate systems:** A raster and polygon may look aligned in code but still be using different coordinate systems. Confirm the two layers share the same coordinate system before comparing them.
- **Forgetting to project before measuring area:** Area in longitude/latitude is not a true metric area. Reproject raster and vector layers before estimating flood extent.
- **Using a low-resolution DEM for a very small coastline:** Coastal flooding often depends on subtle elevation differences. If the raster is too coarse, small dunes and low passes may be missed.
- **Treating a DEM as if it were a water map:** A DEM shows elevation, not water. It tells us where water might collect, but it does not show actual flooding by itself.
- **Over-reading a single map:** A map is evidence, not proof by itself; weigh resolution, data source, and assumptions.

- **Assuming that a good-looking map means correct analysis:** A map can look visually correct even when the coordinate systems, resolution, or layer alignment are wrong. Always check the data source, coordinate system, and spatial units before interpreting results.

3.6 What Makes a Map Scientifically Meaningful

- Spatial data links information to location.
- Vector data uses points, lines, and polygons.
- Raster data uses grids of values.
- Coordinate systems must match before layers are compared.
- Elevation rasters are ideal for introductory geoscience analysis.
- DEM-based analysis can help us understand how terrain influences environmental processes.

> Where These Methods Lead
>
> In this chapter you turned point measurements into maps and filled the space between known locations. Researchers use the same idea to watch the ground itself move.
> Parag and colleagues (2025) used R to study how fast the land is sinking under Long Island, New York. At each of twelve continuous GPS stations they measured one long-term sinking rate, then used spatial interpolation, the same map-filling step you met here, to turn those few rates into a single map. The map showed sinking of about two millimeters a year near New York City, with a few areas rising instead. Tracking this slow motion matters for a low, coastal place, because it helps planners manage groundwater and design infrastructure for a surface that is gradually dropping as the sea rises.
> The step from scattered points to a continuous map is one you have now done yourself.

3.7 From Analysis to Poster

A good scientific map for a poster should be:

- simple,
- readable,
- labeled clearly,
- based on one main message,
- supported by a legend and scale,
- easy to interpret without long explanation.

For this chapter, strong poster figures would be:

1. New York State boundary map.

2. New York elevation raster map.
3. Long Island study area map.
4. DEM clipping workflow for a chosen study area.
5. Hillshade or elevation histogram to show terrain structure.

Keep the map clean. Let the pattern stand out on its own.

3.7.1 Exercises

1. Change the New York State DEM zoom level and compare the result.
2. Transform the Long Island boundary to a different projected coordinate system and inspect how the coordinate units change.
3. Clip a DEM to a different study area and compare the result.
4. Create a hillshade for the New York State DEM and compare it to the original DEM.

3.7.2 References (Data and Further Reading)

1. Bivand, R. S., Pebesma, E., & Gomez-Rubio, V. *Applied Spatial Data Analysis with R.* https://asdar-book.org

2. Pebesma, E. *Simple Features for R.* https://r-spatial.github.io/sf

3. Hijmans, R. J. *terra: Spatial Data Analysis.* https://rspatial.github.io/terra

4. U.S. Census Bureau TIGER/Line Shapefiles. https://www.census.gov/geographies/mapping-files/time-series/geo/tiger-line-file.html

5. elevatr package documentation. https://cran.r-project.org/package=elevatr

6. Lovelace, R., Nowosad, J., & Muenchow, J. *Geocomputation with R.* https://r.geocompx.org

7. Wickham, H., Çetinkaya-Rundel, M., & Grolemund, G. *R for Data Science.* https://r4ds.hadley.nz

8. Natural Earth Map Data. https://www.naturalearthdata.com

9. EPSG Coordinate System Database. https://epsg.io

10. maps package documentation. https://cran.r-project.org/package=maps

11. tigris package documentation. https://cran.r-project.org/package=tigris

12. Parag, A. S., Marsellos, A. E., & Tsakiri, K. G. (2025). *Geostatistical Analysis of Long-Term Subsidence Effects on Isostasy of Long Island Using Continuous GPS Displacement Rates Using R.* In 32nd Conference on the Geology of Long Island and Metropolitan New York, April 2025.

13. Maliszka, M., Johnson, A., Darnaud, D., & Marsellos, A. E. (2020). *An Approach to Understanding the Landscape Exposure Rate for the Post-Wisconsin Late Stage Glacial Melting on Long Island, New York Using a Glacial Withdrawal Simulation.* In 27th Conference on the Geology of Long Island and Metropolitan New York, April 2020.

14. Appel, C., Rahman, M. M., Boddu, D., & Marsellos, A. E. (2024). *Comparing the Effects of Coastal Erosion on Unprotected vs. Protected Shorelines Using GIS and 2011/2014 LiDAR Data.* In 31st Conference on the Geology of Long Island and Metropolitan New York, April 2024.

15. Runkel, M., & Marsellos, A. E. (2026). *GIS-Based Flood Risk Assessment of Fire Island, NY: A LiDAR-Derived DEM Analysis from Robert Moses to Watch Hill.* In 33rd Annual Conference on the Geology of Long Island and Metropolitan New York, April 2026.

Chapter 4

Time Series Fundamentals

Understanding Trends and Cycles in Data

4.1 Introduction

A **time series** is a sequence of observations collected over time. In Earth, environmental, and sustainability sciences, time series data are everywhere: temperature records, river discharge, earthquake occurrences, atmospheric CO_2 levels, and GPS displacement measurements. Time series analysis is widely used in sustainability research to study climate variability, energy demand, emissions, groundwater change, ecosystem behavior, and environmental monitoring systems.

In Chapter 2, we prepared datasets so they could be trusted and analyzed correctly. Now we move to the next step: understanding how data change through time.

The goal is not just to draw a line plot, but to recognize patterns, trends, cycles, and relationships that help us understand Earth processes.

> **i** The main question
>
> Time series analysis asks one main question:
> "How does a variable change through time?"

Understanding how these values change over time allows us to:

- detect long-term changes, such as climate warming or groundwater decline
- identify repeating patterns, such as seasonal cycles
- recognize unusual events, such as extreme weather or seismic activity

Unlike regular datasets, time series data are **ordered**, and this order carries meaning. Ignoring the time dimension can lead to incorrect conclusions.

What you will be able to do

By the end of this chapter, you will be able to:

- plot a time series as a line over time and describe what you see
- identify the three parts of a time series: trend, seasonality, and noise
- smooth a noisy series with a moving average to see the broader pattern
- read an autocorrelation plot to see how a series relates to its own past values (lags)
- use cross-correlation to compare two time series and see which one leads or lags
- use a periodogram to find the strongest repeating cycle (such as a 12-month season)

Why Order Matters

A time series can be viewed as a sequence of events unfolding over time, where order matters just as much as the values themselves.

4.2 Visualization of Time Series

The first and most important step is always **visualization**. A simple line plot often shows patterns such as trends and cycles, and unusual changes more clearly than any numerical summary.

4.2.1 Visualizing Change Through Time

In time series analysis, the order of observations matters. A groundwater value from 2020 should not appear before a value from 2010. Time series plots help us see whether values rise, fall, repeat, or suddenly change over time.

This code creates the made-up ("simulated") dataset we use throughout this chapter; run this block first before trying any other example. You do not need to understand every symbol yet: `seq.Date` builds the monthly dates, the `sin(...)` line creates a realistic temperature (a slow rise, plus a yearly up-and-down cycle, plus a little randomness), and `data.frame` puts the dates and temperatures into a table.

```
# Example: Simulated temperature data
set.seed(1)
time <- seq.Date(from = as.Date("2020-01-01"),
                 by = "month", length.out = 60)
# rnorm(60, 0, 1) draws 60 small random numbers to mimic real-world noise
temperature <- 10 + 0.05 * (1:60) + 5 *
```

```
  sin(2 * pi * (1:60) / 12) + rnorm(60, 0, 1)

temp <- data.frame(time, temperature)

plot(
  temp$time, temp$temperature, type = "l", lwd = 2,
  main = "Temperature Time Series",
  xlab = "Time", ylab = "Temperature (°C)",
  cex.main = 0.7
  # cex.lab  = 0.8,
  # cex.axis = 0.8
)
grid()
```

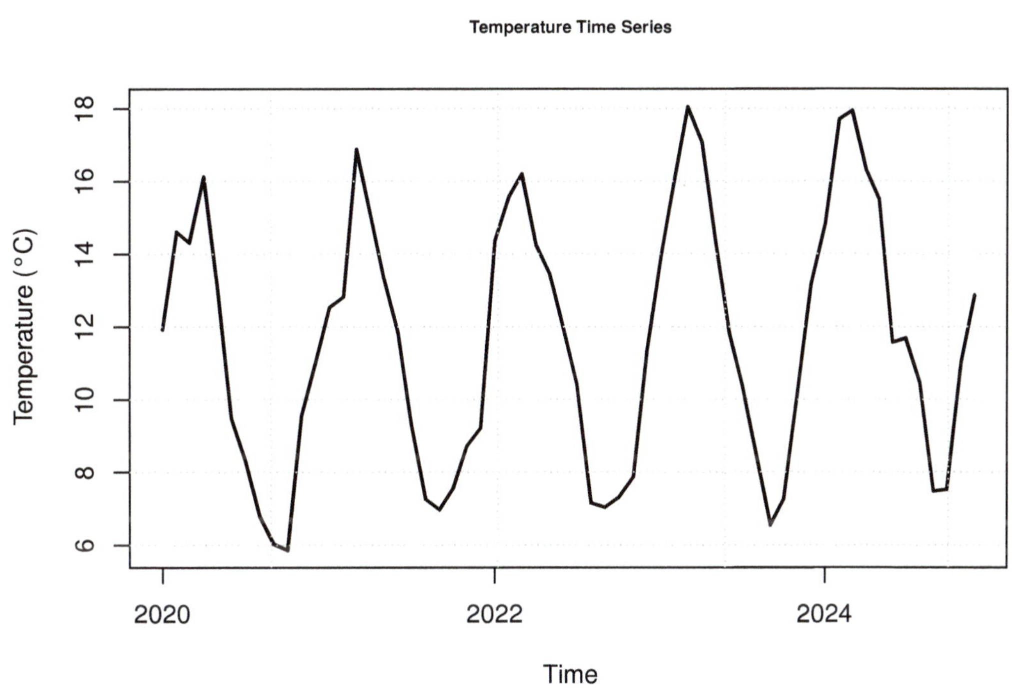

Figure 4.1: Simulated temperature time series showing trend, seasonality, and noise.

The `set.seed(1)` line makes the random numbers reproducible. Think of it like shuffling a deck of cards in exactly the same way every time. That deck still looks random, but it produces the same sequence whenever you start from seed 1. Without set.seed(), R would shuffle differently each time, and your numbers would be slightly different from the ones shown in this book. The number 1 is just a label for the starting point. Any whole number works (such as 42 or 2026); what matters is that everyone uses the same one.

We named the data frame (a table of rows and columns) `temp` because it contains temperature data. As you work through the book, try to choose short but meaningful names that describe the contents of an object. Clear names make code easier to read and understand later. Avoid generic names such as `data` or `df`, which do not indicate what the object actually stores.

> **Ask**
>
> When you first inspect a time series, ask four questions:
>
> - Is there a trend?
> - Is there seasonality?
> - Is there noise?
> - Are there unusual events or sudden jumps?
>
> Noise means random variation that does not follow a clear pattern.

4.2.2 Identifying Patterns

When looking at a time series, we usually ask simple questions first:

- Is the variable generally increasing or decreasing?
- Does a pattern repeat regularly?
- Are there sudden spikes or unusual events?
- Does the series look smooth or noisy?

These visual questions often help us recognize important behavior before any formal analysis begins.

> **Start with the Plot**
>
> Always look at the raw plot before applying models or statistics.
> Many problems become obvious visually before they appear mathematically.

The series in Figure 4.1 illustrates three common features that appear in many time series:

- **Trend**: an overall increase or decrease
- **Seasonality**: a repeating pattern with a regular interval
- **Noise**: short-term random variation

Noise (sometimes called residual variation) represents short-term fluctuations that do not follow a clear pattern.

⚠ Not Every Pattern Is Meaningful

Do not interpret every small up-and-down as a meaningful signal. In time series data, some variation is simply random noise.

4.3 Components of a Time Series

These recurring patterns can usually be grouped into three major behaviors:

- **Trend** → long-term direction, such as gradual warming or cooling
- **Seasonality** → repeating patterns, such as yearly temperature cycles
- **Noise** → random fluctuations that do not follow a clear pattern

Simple analogy

Think of ocean waves:

- Trend = rising sea level
- Seasonality = tides
- Noise = small ripples

or temperature:

- a long-term warming trend,
- repeating summer/winter cycles,
- and day-to-day weather variability.

Components Often Overlap

In real environmental data, these components often overlap. A time series may have both trend and seasonality at the same time.

In one published study, the same kind of decomposition was used to split a river's daily flow into long-term, seasonal, and short-term parts; modeling each part separately explained about 80 percent of the flow in Schoharie Creek, New York (Tsakiri et al., 2014).

4.4 Smoothing

Real-world measurements are often noisy. Short-term fluctuations can make larger patterns difficult to see. Smoothing reduces this short-term variation so the broader pattern becomes easier to see. One of the simplest smoothing tools is the **moving average**, explained next.

4.4.1 Moving Average

A moving average works like a sliding window. Instead of looking at one value at a time, we calculate the average of nearby values. This reduces short-term noise and helps reveal broader patterns.

```
library(zoo)

# rollmean() = rolling (moving) average; fill = NA leaves blanks at the ends
# where a full 5-value window isn't available
temp$smoothed <- rollmean(temp$temperature, k = 5, fill = NA, align = "center")

plot(
  temp$time, temp$temperature, type = "l", lwd = 1,
  main = "Temperature Time Series",
  xlab = "Time", ylab = "Temperature (°C)"
)
lines(temp$time, temp$smoothed, col = "red", lwd = 2)
grid()
```

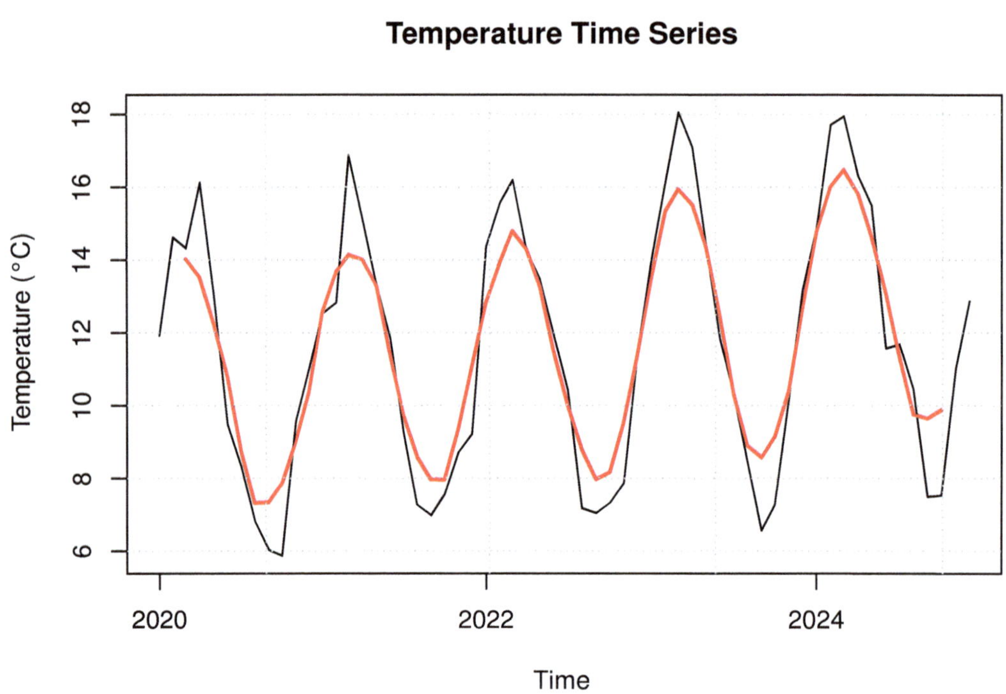

Figure 4.2: Time series with moving average smoothing (red) overlaid on raw data.

Here, `k = 5` means that each smoothed value is calculated using five neighboring observations. Larger values of k produce a smoother curve, while smaller values preserve more short-term variation.

> **ℹ Smoothing Can Hide Details**
>
> The smoothed line in Figure 4.2 shows the broader structure of the data, but too much smoothing can hide important details. Always compare the smoothed version with the original data to ensure you are not losing critical information.

4.4.2 Comparing the Original and Smoothed Series

Use smoothing when you want to compare the raw data with the overall pattern.

- Raw data show short-term fluctuations and noise
- Smoothed data highlight broader patterns and trends

> **⚠ Keep the Original Data**
>
> Always keep the original data visible when possible. The goal is not to replace the original data, but to compare the noisy series with the smoothed version and better understand the underlying trend.

4.5 Stable vs Changing Behavior (Stationarity)

Some time series behave relatively consistently through time, while others change strongly in their average level or variability. Series with strong trends or changing variability are harder to model because their behavior is not stable over time. A time series is **stationary** if its overall behavior stays roughly the same over time. Picture a calm lake whose level moves up and down but never rises steadily or becomes more extreme. In simple terms, the average level and the size of the variations do not drift strongly as time passes. Non-stationary series show strong trends, changing variability, or other systematic changes through time.

Why this matters:

- many time series methods work best when the data are stationary
- non-stationary data can produce misleading patterns or unreliable models

Simple idea:

- stationary = stable behavior
- non-stationary = changing behavior

> **i What is stationarity?**
>
> In simple terms, stationarity means that the overall behavior of the series stays relatively stable through time. A stationary series does not show a strong overall trend or changing variability over time. In more formal terms, the average level (mean) and variability (variance) remain roughly constant.

> **! Always check for trends**
>
> A strong trend often makes a series non-stationary. Before using more advanced methods, always ask whether the data are changing over time in a systematic way.

For those who would like to go a step further:

To check stationarity more formally, you can use the Augmented Dickey-Fuller test from the tseries package:

```
# install.packages("tseries") # Uncomment if needed
library(tseries)
adf.test(temp$temperature)
```

```

	Augmented Dickey-Fuller Test

data:  temp$temperature
Dickey-Fuller = -8.3101, Lag order = 3, p-value = 0.01
alternative hypothesis: stationary
```

The test reports a *p-value*, a number between 0 and 1 that tells you how surprising the result is: small means "strong evidence." Look at the `p-value` line in the printed output; you can ignore the other numbers for now. A p-value below 0.05 suggests the series is stationary. A p-value above 0.05 suggests a non-stationary series that may need *detrending* (removing the overall upward or downward drift) before further analysis. For the simulated temperature series used here, the test returns a small p-value (about 0.01), so it reports the series as **stationary**. This can seem surprising, because the series clearly drifts upward. The key is the *kind* of trend, not just whether you can see one. When you zoom in, almost any real series shows some variation or slope; that alone does not break stationarity. What this test looks for is whether the series wanders off unpredictably and never settles (the behavior of a *random walk*), which would make it non-stationary. Our series climbs along a steady, predictable line rather than wandering, so apart from that regular trend it is stable, and the test reports it as stationary.

So far, we examined how a series changes overall through time. Now we ask a different question: "How strongly are nearby observations related to each other?"

4.6 Autocorrelation

Many Earth variables "remember" their previous state.

For example: - today's temperature is often similar to yesterday's, - groundwater levels change gradually, - climate variables often evolve smoothly through time.

This persistence through time is called autocorrelation. **Autocorrelation** compares **one** time series with itself. It measures how a time series relates to itself at different time *lags* (a lag is simply a step backward in time, so "lag 1" means comparing each value with the one just before it). It helps answer questions such as: Does today look like yesterday (1-day lag)? Does this month resemble the same month last year (12-month lag)?

> **i What is autocorrelation?**
>
> Autocorrelation measures how similar a time series is to itself at different time delays (called **lags**).

> **Think Before Coding**
>
> - If the data have seasonality, at what lag would you expect peaks?
> - Do you expect strong short-term dependence?
> - How far back might the system "remember"?
>
> Try to anticipate the autocorrelation pattern.

```r
# acf() computes and plots the AutoCorrelation Function (ACF)
acf(temp$temperature, main = "Autocorrelation Function")
```

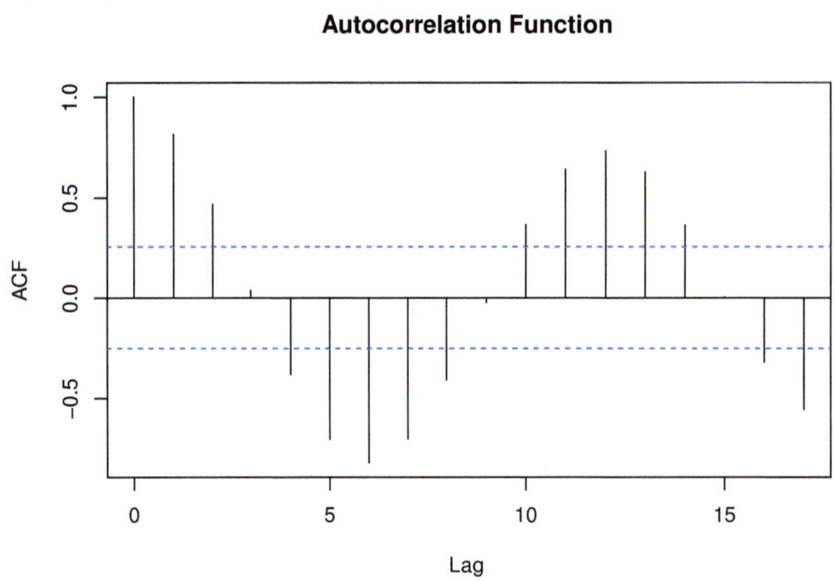

Figure 4.3: Autocorrelation function showing dependence of the series on past values.

i How to read the ACF plot

- The horizontal axis is the **lag**: how many months back you are comparing (0, 1, 2, ...).
- The vertical axis is the **correlation**, from -1 to +1. The bar at lag 0 is always 1, because a series compared with itself at no time shift is identical.

The dashed blue horizontal lines on the ACF plot mark the boundary of statistical significance. Spikes that extend beyond those lines are considered statistically significant, meaning the relationship at that lag is unlikely to be due to chance alone.

Interpretation: Strong short-lag peaks indicate that nearby observations remain similar through time (Figure 4.3). The plot shows that temperatures close together in time are more similar than temperatures farther apart in time, especially at short lags (1 to 2 months). Such behavior is similar to an echo, where current values reflect past conditions. In this sense, today's temperature "remembers" the past. Strong spikes at 1 to 2 months mean recent influence, while a small rise near 12 months, if visible, reflects a yearly seasonal cycle (same month each year).

- high values at certain lags suggest repeating behavior in the data
- seasonal data often show peaks at regular intervals
- persistent correlation can indicate memory in the system

Correlation Is Not Causation

Autocorrelation does not prove a cause. It only shows that values near each other in time are related in some way but not necessarily causing one another.

Autocorrelation examines relationships within one series. We now extend that idea by comparing two different time series together.

4.7 Cross-Correlation

Autocorrelation compares a series with itself. **Cross-correlation** compares **two** different time series to see whether they change together, and helps identify which variable tends to lead or lag another.

! Before Interpreting Correlation

- Do you expect one variable to lead the other?
- Could both variables be responding to the same process?
- What lag would make physical sense?

Be careful: correlation does not imply causation.

For example, precipitation may influence groundwater levels after a delay. Cross-correlation helps us explore whether changes in one variable are followed by changes in another.

```
# Simulate a second variable (e.g., humidity)
temp$humid <- 50 + 10 * sin(2 * pi * (1:60) / 12 + pi / 4) + rnorm(60, 0, 3)

plot(temp$time, temp$humid, type = 'l',
     main = "Humidity Time Series",
     xlab = "Time",
     ylab = "Humidity (%)")
grid()
```

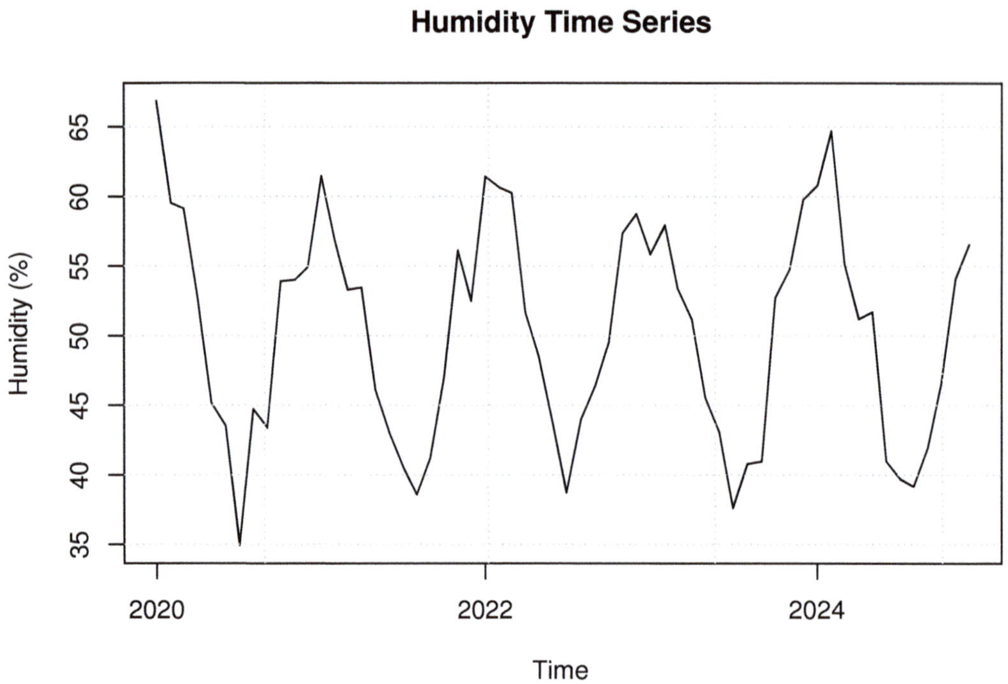

Figure 4.4: Simulated humidity time series showing a seasonal pattern similar to temperature.

The plot (Figure 4.4) shows a second time series (humidity) that has a similar seasonal pattern to temperature. This prepares us for exploring their relationship through cross-correlation.

```
# ccf() computes and plots the Cross-Correlation Function (CCF) between two
# series
ccf(temp$temperature, temp$humid, main = "Cross-Correlation")
```

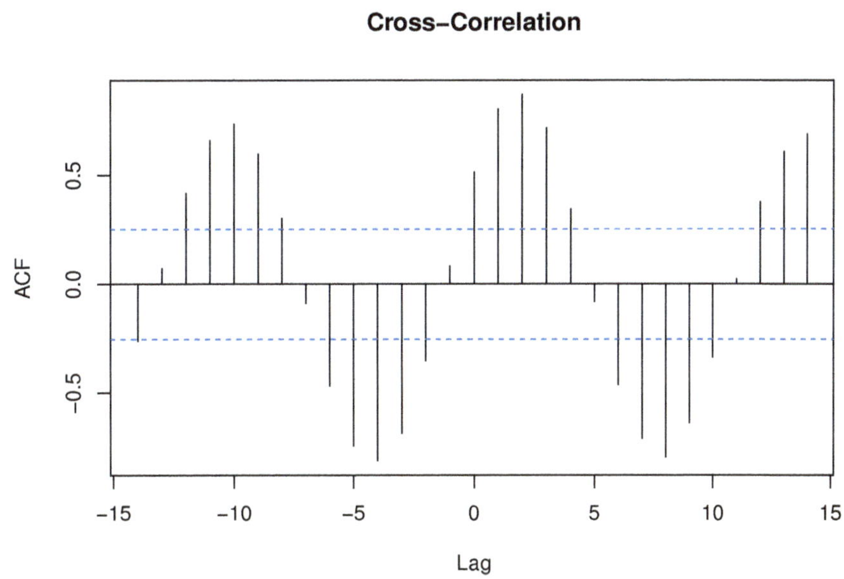

Figure 4.5: Cross-correlation between temperature and humidity showing lag relationships.

> **i** How to read the CCF plot
>
> - The horizontal axis is the **lag**, running from negative on the left, through 0 in the middle, to positive on the right (lag 0 means the two series move at the same time).
> - The vertical axis is the **correlation**, from -1 to +1. The tallest bar marks the lag where the two series line up best.

- **Negative lags** (left side of the plot) suggest that temperature changes occur before humidity changes (temperature leads)
- **Positive lags** (right side of the plot) suggest that humidity changes occur before temperature changes (humidity leads)

Interpretation: The plot (Figure 4.5) shows that temperature and humidity share a strong repeating seasonal pattern. Peaks at multiple lags occur because both variables follow the same annual cycle rather than because one variable directly causes the other. This is an important reminder that cross-correlation can reflect shared seasonality instead of a true physical lag relationship.

This can be understood as two variables responding to the same underlying cycle, like two dancers following the same music. Sometimes one appears to move first, sometimes the other.

- peaks suggest correlation at specific lags
- useful for exploring lagged relationships, such as rainfall and river discharge
- helps show whether one series responds before or after another

> ⚠ Correlation does not imply causation
>
> Cross-correlation does not establish causation. A lag relationship may reflect a shared process rather than direct influence.

> **i** Shared cycles can be misleading
>
> Humidity appears to lead temperature by about 2 months, but this is due to both variables following the same seasonal cycle rather than a true physical relationship.

Autocorrelation and cross-correlation help us study relationships through time. Another important question is whether a repeating cycle exists inside the data.

4.8 Repeating Patterns Through Time

So far, we explored patterns by following the data through time. Now we look at the same data a different way: by asking which repeating cycles they are built from (a "frequency" view).

Many geoscience processes repeat in cycles. These cycles may be annual, monthly, daily, or more complex. In time series analysis, repeating cycles are often described using terms such as period, frequency, and periodicity.

Examples include:

- seasonal temperature cycles
- tidal cycles
- climate oscillations

A yearly seasonal cycle has a **period** of one year because the pattern repeats every year. The **period** is how long it takes for a repeating pattern to complete one full cycle, and **periodicity** simply means that a pattern repeats at a regular interval. For example, a 12-month period means the pattern repeats once every year. A **periodogram** is a chart that helps identify which repeating cycles are strongest in the data, and a strong peak at a specific period indicates that the data have a strong repeating pattern at that interval.

> **!** Interpreting the Pattern
>
> - Do you expect a repeating cycle?
> - What period might be dominant (e.g., 12 months)?
> - Will multiple cycles be present?
>
> Think in terms of cycles before looking at frequencies.

A periodogram helps us identify repeating cycles in a time series. Instead of examining the data directly through time, a periodogram searches for repeating cycles inside the series. It measures cycles as a frequency (how many cycles per month), and we use **1/freq** to convert that into a period (the number of months per cycle), which is easier to read.

For example, frequency analysis has confirmed the expected 365-day yearly cycle in New York City temperature records, alongside the sun's roughly 11-year sunspot cycle (Colucci and Marsellos, 2025).

To compare temperature and humidity cycles more easily, we display the two plots side by side. Visual comparison is often the first step before applying more advanced methods to study relationships between time series.

```
library(TSA)
# look at which repeating cycles (frequencies) are present in the data
par(mfrow = c(1,2)) # set up side-by-side plots
# periodogram(..., plot = FALSE) returns $freq (cycles per month) and $spec
# (cycle strength);
# we plot 1/$freq so the x-axis shows the period in months
sp_t <- TSA::periodogram(temp$temperature, plot = FALSE)
plot(1/sp_t$freq,sp_t$spec, type = 'l',
     xlab = "Period (months)",
     ylab = "Spectral Density",
     main = "Periodogram of Temperature")

sp_h <- TSA::periodogram(temp$humid, plot = FALSE)
plot(1/sp_h$freq,sp_h$spec, type = 'l',
     xlab = "Period (months)",
     ylab = "Spectral Density",
     main = "Periodogram of Humidity")
par(mfrow = c(1,1)) # restore default layout
```

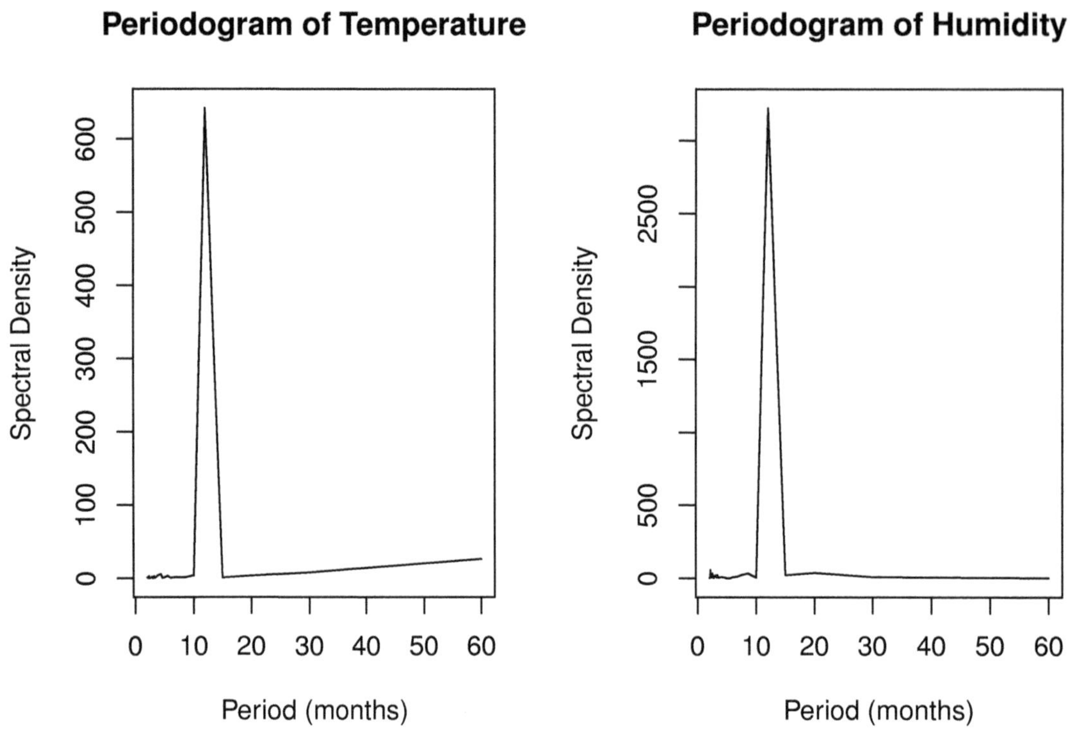

Figure 4.6: Periodograms of temperature and humidity showing dominant cycles.

Interpretation: The plot (Figure 4.6) shows the dominant periodic signals in both series. On these charts the height of the curve ("spectral density") shows how strong each cycle is, so a tall peak means a strong repeating cycle at that period. The periodogram shows a dominant 12-month cycle in both temperature and humidity. This shared seasonality explains their strong cross-correlation. Both fluctuate with the same seasonal cycle rather than directly controlling one another.

- peaks in a periodogram represent dominant periodic signals
- a strong peak often corresponds to a repeating cycle, such as annual seasonality
- no smoothing is applied to the periodogram, keeping the interpretation straightforward

If you see a strong peak corresponding to a 12-month cycle, this is a clear indication of seasonality in the data.

4.9 R Code Examples (Workflow Summary)

A simple workflow for time series analysis is:

1. load the data
2. plot the time series
3. identify trend, seasonality, and noise

4. smooth the series if needed
5. check autocorrelation
6. explore periodicity
7. build a simple model if appropriate (introduced later)

```r
summary(temp)
```

```
      time                temperature        smoothed           humid      
 Min.   :2020-01-01   Min.   : 5.864   Min.   : 7.303   Min.   :34.93  
 1st Qu.:2021-03-24   1st Qu.: 8.404   1st Qu.: 9.306   1st Qu.:43.47  
 Median :2022-06-16   Median :11.763   Median :11.350   Median :51.69  
 Mean   :2022-06-16   Mean   :11.633   Mean   :11.594   Mean   :50.34  
 3rd Qu.:2023-09-08   3rd Qu.:14.374   3rd Qu.:13.939   3rd Qu.:56.23  
 Max.   :2024-12-01   Max.   :18.050   Max.   :16.461   Max.   :66.86  
                                       NA's   :4                       
```

Keep it simple

A clean workflow is often more valuable than a complicated one. In introductory analysis, clarity is more important than trying every available method.

The following example combines several ideas from this chapter:

- visualization,
- smoothing,
- pattern recognition,
- and interpretation.

Instead of studying methods separately, we now apply them together to a single example series.

4.10 Example / Application

4.10.1 Case Study: Temperature Trends

In this example, we use a monthly temperature series that contains a gradual upward trend and a clear seasonal cycle.

Possible observations:

- a rising trend may suggest warming
- the seasonal cycle may reflect annual solar variation
- short-term fluctuations may represent weather variability

```
plot(temp$time, temp$temperature, type = "l",
     main = "Monthly Temperature Series",
     xlab = "Time", ylab = "Temperature (°C)")
```

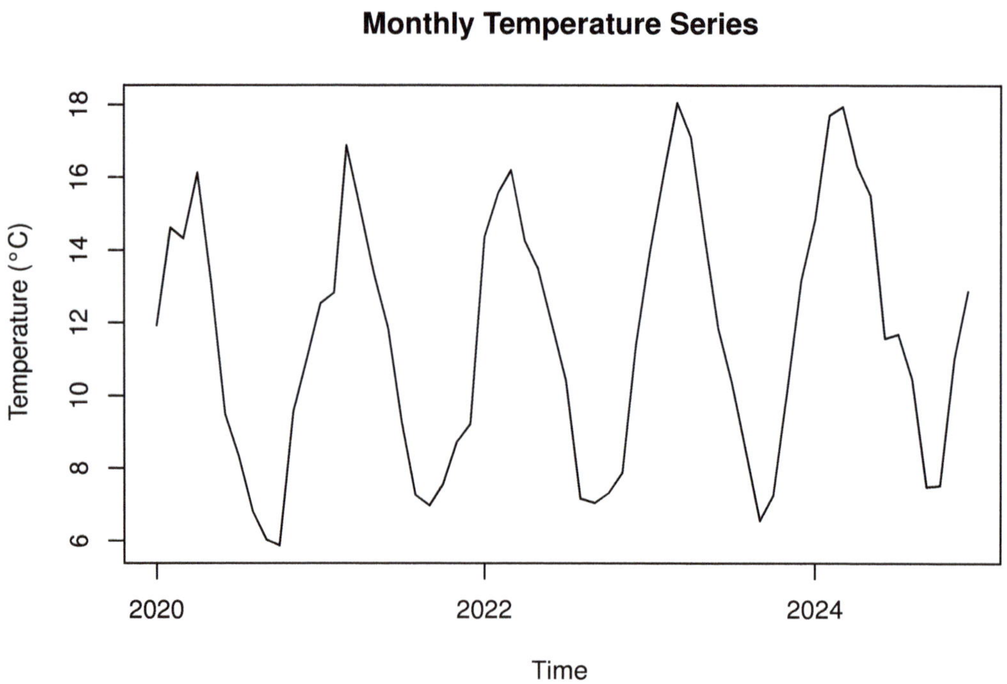

Figure 4.7: Monthly temperature time series showing trend and seasonal variation.

The series in Figure 4.7 illustrates a typical environmental time series with both trend and seasonality.

Connect Patterns to Processes

In geoscience, the most important question is not only "What pattern do we see?" but also "What physical process might explain it?"

4.11 Interpretation of Results

From a geoscience perspective:

- trend may indicate climate change or long-term environmental shifts
- seasonality reflects Earth system processes
- autocorrelation suggests memory in the system
- cross-correlation may indicate interactions between variables

> ⚠ Statistical Patterns Need Scientific Context
>
> A statistical pattern does not automatically prove a physical cause. Time series analysis helps identify relationships and patterns, but interpretation still requires scientific reasoning. A pattern becomes meaningful only when it also makes scientific sense.

4.12 Common Mistakes to Avoid

- treating time series observations as independent data points
- assuming correlation automatically means causation
- ignoring non-stationarity
- over-smoothing and removing important signals
- interpreting smoothed or interpolated values as direct measurements
- confusing noise with meaningful patterns

> ⚠ A Good Plot Is Not Enough
>
> A visually attractive result is not automatically a correct result. Always check whether the method matches the data.

A common mistake is to see a correlation between two time series and assume that one causes the other. In reality, both series may be responding to a shared underlying process (e.g., seasonality) rather than directly influencing each other. Strong peaks in cross-correlation do not mean that one variable causes the other. If both time series share a similar seasonal pattern, they can appear strongly related at multiple lags, even suggesting that each one "leads" the other. Always check whether the relationship is driven by a common underlying cycle rather than a true cause-and-effect connection.

4.13 What Time Series Can and Cannot Tell Us

- time series data are ordered and dependent
- visualization is the most important first step
- the main components are trend, seasonality, and noise
- smoothing can reveal structure, but it can also hide detail
- autocorrelation shows internal dependence in a series
- cross-correlation helps compare two time series across lags
- repeating patterns do not automatically imply causation
- interpretation should be consistent with the underlying Earth processes

In this chapter, we focused on understanding patterns through time. In the next chapter, we move from describing patterns to building models that can explain relationships and make predictions.

> **Where These Methods Lead**
>
> In this chapter you split a time series into a long-term trend, a repeating seasonal part, and short-term changes, and you used a periodogram to find hidden cycles. Both ideas lead straight into research.
> Tsakiri and colleagues (2014) split the daily flow of Schoharie Creek in New York into those same three parts and built a separate model for each. Together the parts explained about 80 percent of the river's flow, far more than a single model of the raw series, and a model like this can warn of dangerous high flows before they arrive, which helps communities prepare for floods.
> Searching for cycles leads somewhere too. Colucci and Marsellos (2025) studied New York City temperature records back to 1948 and confirmed the strong yearly (365-day) cycle in the temperature, alongside the sun's well-known eleven-year sunspot cycle in solar activity, an early step toward asking whether the two are connected.
> You have now used both yourself: separating a series into its parts, and searching it for repeating cycles.

4.14 From Analysis to Poster

A strong time series figure should include:

- a clear time axis
- labeled units
- raw data and smoothed trend, if appropriate
- annotations for key events or changes

Good practice:

- avoid clutter
- use readable labels
- explain what the pattern means scientifically

> One Clear Figure Is Better
>
> For a poster, one strong figure with a clear message is better than several crowded plots.

Time series analysis is not only about detecting mathematical patterns. It is about connecting those patterns to real Earth processes and asking whether the interpretation makes physical sense.

4.15 Questions to Consider Before Continuing

Before moving to the exercises, ask yourself:

- Which patterns are visible directly in the time series and which only became visible after analysis?
- Could any apparent pattern be an artifact of smoothing, filtering, or data collection?
- Which Earth processes might explain the patterns observed?
- If the same pattern appeared in a different dataset, would you interpret it the same way?

4.16 Exercises

1. Plot a time series of daily temperature data and identify trend and seasonality.
2. Apply a moving average and compare the smoothed result with the original data.
3. Compute and interpret autocorrelation for a real or simulated dataset.
4. Simulate two related time series and explore their cross-correlation.
5. Identify periodic patterns using a periodogram.

4.16.1 References (Data and Further Reading)

1. Hyndman, R. J., & Athanasopoulos, G. *Forecasting: Principles and Practice.* https://otexts.com/fpp3
2. Chatfield, C. *The Analysis of Time Series.*
3. Wilks, D. S. *Statistical Methods in the Atmospheric Sciences.*
4. TSA package documentation. https://cran.r-project.org/package=TSA
5. zoo package documentation. https://cran.r-project.org/package=zoo

6. Tsakiri, K. G., Marsellos, A. E., & Zurbenko, I. G. (2014). *An Efficient Prediction Model for Water Discharge in Schoharie Creek, NY.* Journal of Climatology, 2014, 284137. https://doi.org/10.1155/2014/284137

7. Colucci, D., & Marsellos, A. E. (2025). *Investigation of the Influence of Sunspot Cycles on Temperature Variability in New York City (1948–Present): A Preliminary Wavelet Analysis Approach.* In 32nd Conference on the Geology of Long Island and Metropolitan New York, April 2025.

8. Plitnick, T. A., Lagattolla, R., Tsakiri, K. G., Petrocheilos, C., & Marsellos, A. E. (2019). *Time Series Regression and Spectral Analysis on Nitrate Concentrations in Long Island, New York.* In 26th Conference on the Geology of Long Island and Metropolitan New York, April 2019.

9. Marsellos, A. E., Tsakiri, K. G., & Tentomas, G. (2022). *A Geospatial Analysis of the Sunspots Effect on Groundwater Levels in Long Island, NY During 2002-2021.* In 29th Conference on the Geology of Long Island and Metropolitan New York, April 2022.

10. Gonzalez, X., Hubbs, D., Dailey, S., Marsellos, A. E., & Tsakiri, K. G. (2024). *Exploring Climate Change Effects on Sinkhole Formation: Long-Term Temperature Analysis in Long Island, NY (1948-2024).* In 31st Conference on the Geology of Long Island and Metropolitan New York, April 2024.

Chapter 5

Modeling, Prediction, and Machine Learning

Building and Interpreting Predictive Models

5.1 Introduction

The previous chapters introduced individual tools and ideas:

- how to work with data
- how to analyze spatial patterns
- how to analyze time series

In this chapter, we bring those ideas together using real Earth and sustainability applications.

Earth systems are rarely controlled by a single factor. A model is a simplified way to describe or predict part of the real world, for example, "hotter days tend to have more ozone." It provides a structured way to represent relationships and patterns in data. Some models help explain how variables are connected, while others help estimate or predict values.

For example, a model may estimate how ozone changes with temperature, how flooding responds to rainfall and terrain, or how groundwater levels respond to rainfall through time. No model captures everything. We should not trust results automatically. A model is an approximation that helps us recognize patterns, make predictions, and test ideas. In Earth and sustainability analytics, models help us understand environmental processes, forecast change, evaluate risk, and support decision-making.

Chapter 4 focused on recognizing patterns through time. Here we begin building models that attempt to describe or predict those patterns quantitatively.

In this chapter, the goal is not mathematical complexity, but interpretation. We will ask:

- What is the model learning?
- Does the result make scientific sense?
- How accurate is the model on new data?
- What are the limits of the result?

Building a model is only part of the process. Scientific interpretation is what gives the model meaning. At this stage, you do not need advanced mathematics. The goal is to learn how to think about relationships in data and how to interpret what a model is telling us.

> **!** Think before coding
>
> Before fitting any model, try to predict the outcome in plain language. Ask yourself what relationship you expect, whether it should be positive or negative, and whether the relationship is likely to be direct or only a correlation.

We begin with simple relationships between variables, then evaluate how models behave, and finally explore introductory prediction and machine learning workflows.

> **i** What you will be able to do
>
> By the end of this chapter, you will be able to:
>
> - fit a simple linear regression line (`lm`) to summarize how one variable relates to another
> - read a fitted model in plain language (its slope, intercept, and R-squared)
> - split data into training and testing sets and measure prediction error with MAE and RMSE
> - build a multiple regression model that uses several predictors at once
> - make a short-term forecast from a repeating seasonal pattern (the seasonal naive method)
> - grow and read a decision tree (a flowchart of yes/no rules) and compare it with linear regression

5.2 Simple Relationships

A simple model often begins with one input variable and one output variable. In statistics, the input is called the predictor (the thing we use to make a guess) and the output is called the response (the thing we are trying to guess). For example, we may want to understand how temperature relates to ozone concentration, or how one environmental variable changes with another.

The goal is not only to predict values. We also want to understand the relationship itself.

For example:

- Does ozone increase when temperature increases?
- Is the relationship strong or weak?
- Does the pattern look clear or noisy?

A model helps us summarize those patterns in a simple way. It also helps us move from simply observing patterns to testing relationships quantitatively.

5.2.1 Linear regression

Linear regression is one of the easiest models to interpret. It fits a straight line through the data so that the line describes the overall trend as clearly as possible. If the slope is positive, the response variable tends to increase as the predictor increases. If the slope is negative, the response variable tends to decrease as the predictor increases.

You can think of the regression line as the "overall direction" of the data cloud. The line does not pass exactly through every observed data point. Instead, it summarizes the general pattern.

This does not mean that one variable causes the other. A regression line only describes association. In geoscience, two variables may move together because they are both influenced by the same process, such as seasonality, climate, topography, or human activity. Correlation is not causation.

> **!** State the Expected Relationship
>
> Before fitting the regression line, describe the expected relationship in one sentence. For example: "I expect ozone to increase with temperature" or "I expect wind to reduce ozone because stronger mixing may disperse pollutants." That prediction gives meaning to the model.

5.2.2 Visualizing relationships

A scatter plot is usually the best first step. It shows the data directly and helps us see whether a straight line is reasonable or whether the relationship is curved, weak, or dominated by outliers.

We will use a built-in dataset in R called `airquality`. It contains daily observations of air quality in New York (May to September 1973), including ozone levels, temperature, wind speed, and solar radiation.

> **i** About the Dataset
>
> You can learn more about this dataset in R using:
> `?airquality`

```r
# A simple environmental example
# Keep only the four columns we need from the airquality dataset.
aq <- na.omit(airquality[, c("Ozone",
                              "Temp",
                              "Wind",
                              "Solar.R")])

plot(aq$Temp, aq$Ozone,
     pch = 19,
     xlab = "Temperature (°F)",
     ylab = "Ozone (ppb)",
     main = "Ozone and Temperature")
grid()

# lm() fits a straight line; read `Ozone ~ Temp` as "explain Ozone using Temp".
lm_simple <- lm(Ozone ~ Temp, data = aq)
abline(lm_simple, lwd = 2)  # add that fitted line to the plot
```

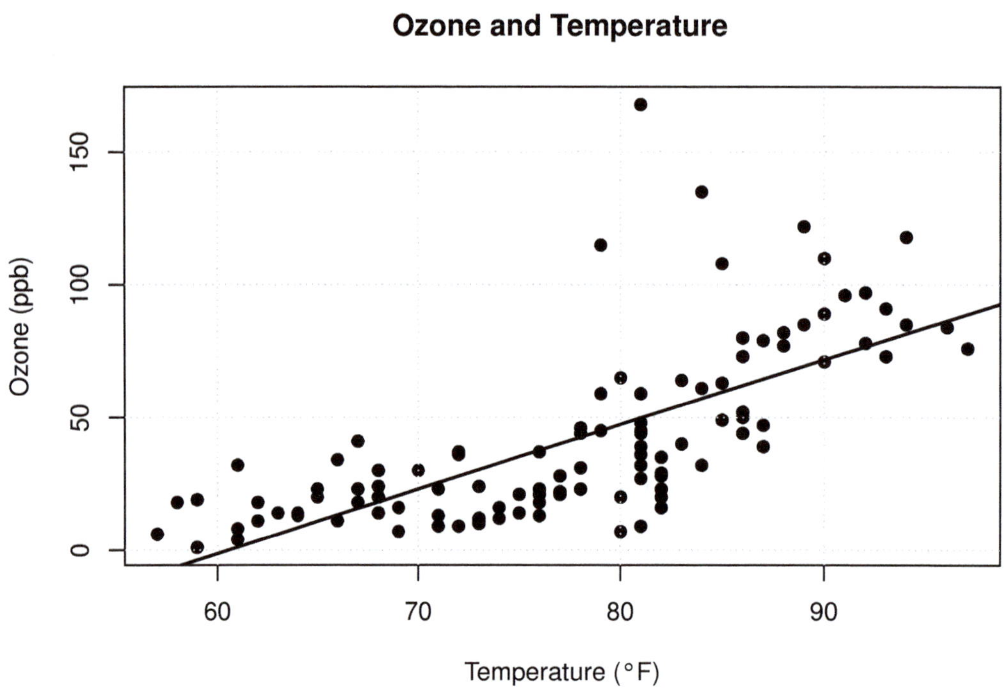

Figure 5.1: Scatter plot of ozone versus temperature with fitted linear regression line.

The relationship between temperature and ozone appears as a scatter of points with a fitted line that provides a simple summary of the overall trend (Figure 5.1). The plot suggests whether a line is reasonable. If the points are widely scattered, the relationship may be weak. If a clear pattern is visible, a linear model can provide a useful first approximation. The line does not capture every detail of the data, but it still provides a useful starting point.

> Always label your axes
>
> A plot is only clear if the reader knows what each axis shows. Give every axis a descriptive label, and include the unit of measurement when one exists (for example, Temperature (°F) or Ozone (ppb)). A bare number on an axis is easy to misread.

> Removing Missing Values
>
> The function `na.omit()` removes rows with missing values (NA).

> Thinking Like an Analyst
>
> Before trusting a relationship, ask:
>
> - Does the relationship make scientific sense?
> - Could another variable influence the result?
> - Is the pattern stable through time?
> - Are we describing a real process, or only a statistical pattern?

5.3 Model Interpretation

Once a model is fit, the next step is to interpret it in plain language. A fitted line has two key numbers. The intercept is the starting value: the predicted response when the predictor is zero (which is not always scientifically meaningful). The slope is the steepness: how much the response changes for each one-unit increase in the predictor.

For example, if temperature has a positive coefficient (its slope number) in an ozone model, the model is saying that ozone tends to increase as temperature increases. That is a statement about association in the available data. It is not proof of direct causation.

Predictions are also important. A prediction is the model's estimate for a new case. When we predict, we are using the pattern learned from past data to estimate an unknown value. That can be useful, but only if the model is appropriate and the new data are similar to the data used to build the model.

5.3.1 Reading the Output

The output below may look complicated at first. Do not try to understand every number yet. It is just a summary table that R prints to describe the fitted line.

Focus only on three ideas:

- the slope (direction of relationship)
- how much of the overall variation the model explains
- and how much prediction error remains

Before focusing on the detailed output, first translate the model into plain language.

```
summary(lm_simple)
```

```
Call:
lm(formula = Ozone ~ Temp, data = aq)

Residuals:
    Min      1Q  Median      3Q     Max 
-40.922 -17.459  -0.874  10.444 118.078 

Coefficients:
             Estimate Std. Error t value Pr(>|t|)    
(Intercept) -147.6461    18.7553  -7.872 2.76e-12 ***
Temp           2.4391     0.2393  10.192  < 2e-16 ***
---
Signif. codes:  0 '***' 0.001 '**' 0.01 '*' 0.05 '.' 0.1 ' ' 1

Residual standard error: 23.92 on 109 degrees of freedom
Multiple R-squared:  0.488, Adjusted R-squared:  0.4833 
F-statistic: 103.9 on 1 and 109 DF,  p-value: < 2.2e-16
```

Where to look in this table

- In the row labeled `Temp`, the `Estimate` column is the slope: about 2.44.
- Near the bottom, `Multiple R-squared` is 0.488 (the 49% mentioned below).
- **`Residual standard error`** is the typical miss, about 23.92 (the ±24 ppb mentioned below).

Reading those numbers from the output above:

- For every 1°F increase in temperature, ozone increases by about **2.4 ppb** on average (this is the slope)
- About **49% of the changes in ozone levels** are explained by temperature in this model, while **51% remains unexplained** (this 49% is the value R labels "R-squared")
- Predictions have an average error of about **±24 ppb**

The intercept is **-147.6**, which represents the predicted ozone when temperature is 0°F. This value is not physically meaningful in this context, and highlights an important limitation: model components do not always have direct real-world interpretation.

> **!** Correlation Is Not Causation
>
> A strong statistical relationship does not imply causation. Temperature and ozone may both be influenced by shared processes such as seasonality, solar radiation, or atmospheric conditions.

```
# Example predictions at selected temperatures
new_temp <- data.frame(Temp = c(60, 75, 90))
predict(lm_simple, newdata = new_temp)
```

```
        1         2         3 
-1.299478 35.287171 71.873819 
```

Predictions are usually more reliable when they stay within the range of the original data (for example, predicting ozone for a 75°F day when our data ran from 57°F to 97°F). This is called **interpolation**. Predicting far beyond the observed range, for example ozone at 120°F, is called **extrapolation** and can become unreliable because we have no data there.

The coefficients should be read carefully. A positive coefficient means the predicted response increases with the predictor. A negative coefficient means the response decreases as the predictor increases. If the coefficient is small or unstable, the relationship may be weak or influenced by other variables.

> **i** Evaluate Beyond the Training Data
>
> A model may look convincing visually but still perform poorly on new data. That is why models should always be evaluated beyond the data used to build them.

5.4 Model Evaluation

A model can look good on the data used to build it and still perform poorly on new data. This is why evaluation matters. We want to know whether the model generalizes, that is, whether it still works on data it has never seen. This is like a student who truly understands a topic and can answer questions that were not on the practice sheet.

The simplest evaluation method is to divide the data into training and testing sets.

- **training data** → used to build the model
- **testing data** → used to evaluate predictions on unseen data

We fit the model on the training data, then evaluate its predictions on the testing data.

To evaluate predictions, we compare the predicted values with the observed values. The difference between them is called the **prediction error**. Every model makes mistakes, so evaluation helps us measure how large those mistakes are and whether the model works reasonably well on new data. Two common ways to summarize prediction error are:

- **MAE (mean absolute error)** → average size of the errors
- **RMSE (root mean square error)** → similar to MAE, but gives more weight to large mistakes

Smaller values usually indicate better predictions. However, what counts as "good" still depends on the scientific question and the scale of the data.

We often want a model that is not only accurate, but also understandable and scientifically reasonable. Also, those errors are just numbers. They do not tell us why the model works or whether it captures the true underlying processes.

These metrics are most useful for comparing models. Lower error usually indicates better predictive performance, but there is no single number that guarantees a model is scientifically reliable.

Overfitting happens when a model learns the training data too well, memorizing even the random, meaningless ups and downs ("noise") instead of the real pattern. Like a student who memorizes the practice answers exactly, an overfit model looks impressive on the training set but fails on new cases. A simpler model is often more useful when the goal is interpretation.

> **!** Think before coding
>
> Before calculating error, decide what would count as a useful model. Is the goal to explain the relationship, to predict future data, or to compare competing hypotheses? The evaluation method should match the goal.

5.4.1 Multiple Predictors and Multivariable Models

So far, we used only one predictor: temperature. But environmental systems are usually influenced by many factors at the same time. For example, ozone may also depend on wind, solar radiation, humidity, or seasonal conditions.

A multivariable model allows us to include several variables together instead of relying on only one predictor. Temperature explained about 49% of the changes in ozone levels, meaning that more than half remained unexplained. This suggests that other variables may also influence ozone, which motivates the use of a multivariable model.

To evaluate whether additional variables improve prediction, we now build a model using several predictors together.

The model estimates the relationship between ozone and each variable while accounting for the others at the same time. This helps us understand which variables appear most important in the data.

```
# fixes the random draw so you get
# the same split every time you run this
set.seed(123)
# number of observations (rows)
n <- nrow(aq)
# randomly pick row numbers for 70% of the data to train on
train_id <- sample(seq_len(n),
                   size = floor(0.7 * n))
train <- aq[train_id, ]  # the rows we picked = training dataset
# the minus sign means "all other rows"
# = testing dataset (remaining 30%)
test <- aq[-train_id, ]

# fit model on training data - multivariable regression
model_fit <- lm(Ozone ~ Temp + Wind + Solar.R, data = train)
summary(model_fit) # shows the coefficients and R-squared discussed below
```

```
Call:
lm(formula = Ozone ~ Temp + Wind + Solar.R, data = train)

Residuals:
    Min      1Q  Median      3Q     Max 
-29.912 -16.086  -4.016   7.824  92.987 

Coefficients:
             Estimate Std. Error t value Pr(>|t|)  
(Intercept) -74.18853   28.90334  -2.567   0.0123 *
```

```
Temp          1.84998    0.32151   5.754 1.91e-07 ***
Wind         -3.34508    0.80999  -4.130 9.55e-05 ***
Solar.R       0.04507    0.02867   1.572   0.1202
---
Signif. codes:  0 '***' 0.001 '**' 0.01 '*' 0.05 '.' 0.1 ' ' 1

Residual standard error: 22.4 on 73 degrees of freedom
Multiple R-squared:  0.6131,    Adjusted R-squared:  0.5972
F-statistic: 38.56 on 3 and 73 DF,  p-value: 4.82e-15
```

```
# make predictions on unseen (test) data
model_pred <- predict(model_fit, newdata = test)

# evaluate prediction error
# MAE and RMSE of the model on the test set
mae_model <- mean(abs(test$Ozone - model_pred))
rmse_model <- sqrt(mean((test$Ozone - model_pred)^2))
# print both with labels so each number is clearly named
c(regression_MAE = mae_model, regression_RMSE = rmse_model)
```

```
 regression_MAE regression_RMSE
       15.53134        19.27173
```

Interpreting the Multivariable Results

The earlier simple regression model explained about 49% of the ozone variation using temperature alone. After adding wind and solar radiation, the multivariable model explains about 61% of the variation. This improvement occurs because the model now includes additional environmental variables besides temperature.

Before focusing on the full output, translate the results into plain language.

- Temperature increases ozone by about **1.85 ppb for each 1°F warmer**
- Wind decreases ozone by about **3.3 ppb for each 1 mph stronger** (windier days have less ozone)
- Solar radiation shows a relatively weak relationship in this dataset
- About **61% of the changes in ozone levels** are captured by the model, while **39% remains unexplained**
- The model has an **average prediction error** of about 15 ppb (MAE), with larger errors reflected in an **RMSE** of about 19 ppb

> **!** Not All Variation Is Explained
>
> Even though this model improves prediction compared to a simple regression, it still leaves a large portion unexplained. Real environmental systems are influenced by many interacting processes.

A small error is encouraging, but it does not prove the model is correct. It only tells us that the model worked reasonably well on the data we tested. Models are usually more reliable within conditions similar to the data used to build them.

A **residual plot** is also useful. A residual is simply one prediction's error, how far off a single guess was. Plotting these errors helps us see whether they behave randomly (good) or whether the model is missing an important pattern (a leftover trend in the errors).

```
par(mfrow = c(1, 1))
plot(model_pred, test$Ozone,
     pch = 19,
     xlab = "Predicted Ozone (ppb)",
     ylab = "Observed Ozone (ppb)",
     main = "Observed vs Predicted")
abline(0, 1, lwd = 2)
grid()
```

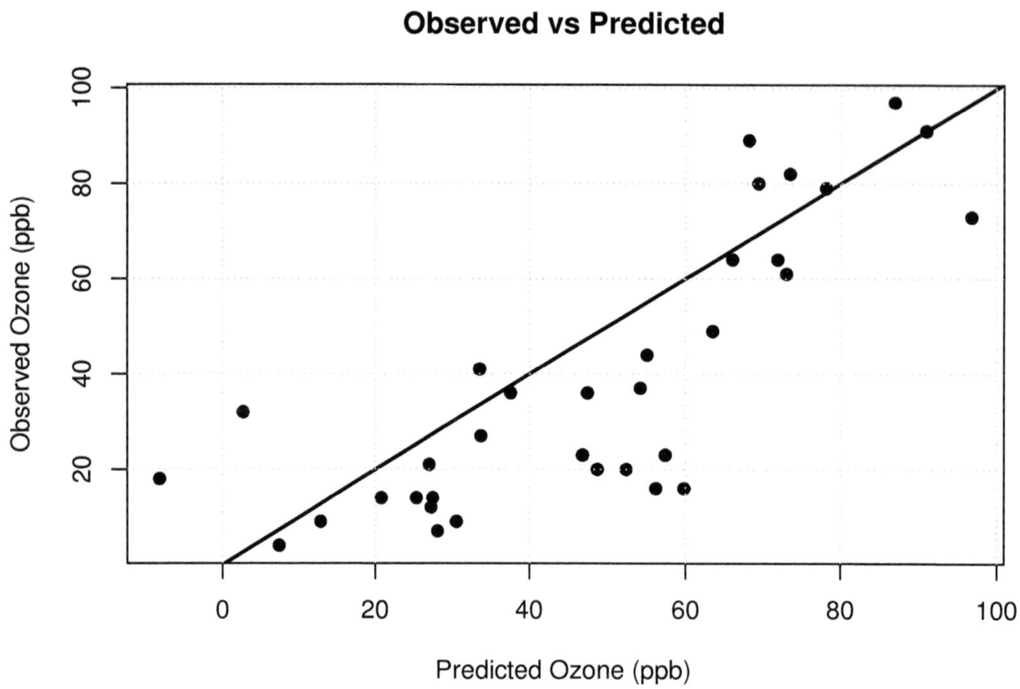

Figure 5.2: Observed versus predicted ozone values from the multivariable regression model.

Points close to the diagonal in Figure 5.2 indicate good agreement between predicted and observed values. However, if points are widely scattered, the model may be missing important information or may not generalize well. The plot should be interpreted in the context of the scientific question and the data.

5.5 Modeling Workflow

A model is only as useful as the workflow around it. Even simple models can produce meaningful results when the question is clear, the model is tested on new data, and the results are interpreted carefully.

A simple and reliable modeling workflow is:

1. Define the problem clearly.
2. Inspect and visualize the data.
3. Choose a model that matches the question.
4. Fit the model on training data.
5. Evaluate the model on new data.
6. Interpret the result in plain language.
7. Check limitations and possible failure points.

This workflow is more important than any single algorithm. A clear question with a simple model is often better than a complicated model used without interpretation.

5.6 Time Series Modeling (Introductory)

Time series data are different from ordinary data because order matters. In a time series, the value at one moment may depend on earlier values. That is why we must be careful when modeling time series.

We should not randomly change the order of the observations, because the sequence through time contains important information. A simple time series model may use the previous value as a forecast, or it may use a clear seasonal pattern. At this stage, the goal is not the most accurate possible forecast. The goal is to understand how time structure influences prediction and how repeating patterns can support sensible short-term forecasts.

5.6.1 Periodicity and repeating structure

Some time series have seasonal cycles or other repeating patterns. A periodogram is a chart that shows which repeating cycle lengths are strongest in the data. Following the style used in earlier chapters, we will use `TSA::periodogram()`.

In this example, we use the built-in `nottem` dataset in R, which contains monthly average air temperatures recorded at Nottingham, UK. The data span from 1920 to 1939, and the temperature is measured in degrees Fahrenheit (°F). The dataset is a classic example of a time series with strong seasonal patterns, making it ideal for demonstrating periodicity and forecasting methods.

```
# Monthly temperature series
# expressed in a simple data frame for plotting
temp_df <- data.frame(
  time = as.numeric(time(nottem)),
  temperature = as.numeric(nottem)
)

plot(temp_df$time, temp_df$temperature,
     type = "l",
     lwd = 2,
     main = "Monthly Temperature Time Series",
     xlab = "Time",
     ylab = "Temperature (°F)")
grid()
```

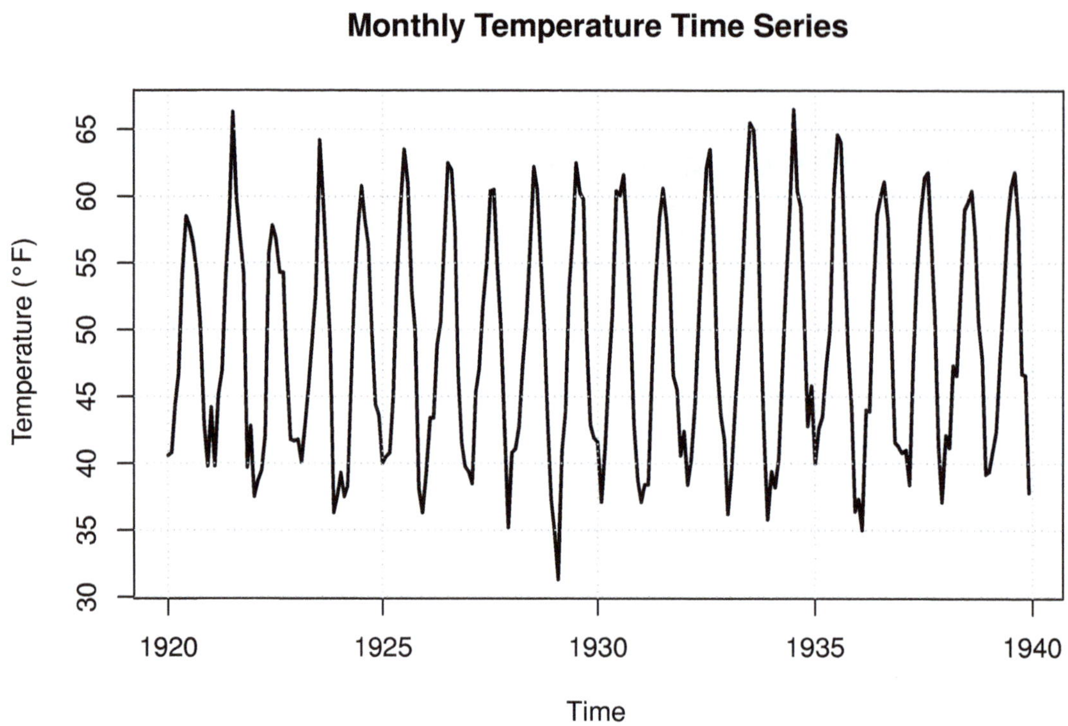

Figure 5.3: Monthly average temperatures from the Nottingham (**nottem**) dataset, showing trend and seasonal variation.

The plot of the monthly temperature series (Figure 5.3) shows a clear repeating pattern, with temperatures rising and falling in a regular cycle. This is a strong indication of seasonality, which is common in many environmental time series. The periodogram helps us visualize which repeating cycles are strongest in the data.

```
# We plot 1/frequency so the x-axis reads as a period in months
# (e.g. 12 = a cycle that repeats once a year); a tall peak = a strong cycle.
sp_t <- TSA::periodogram(nottem, plot = FALSE)
plot(1 / sp_t$freq, sp_t$spec,
     type = "l",
     lwd = 2,
     xlab = "Period",
     ylab = "Spectral Density",
     main = "Periodogram of Monthly Temperature",
     xlim = c(0, 24))
grid()
```

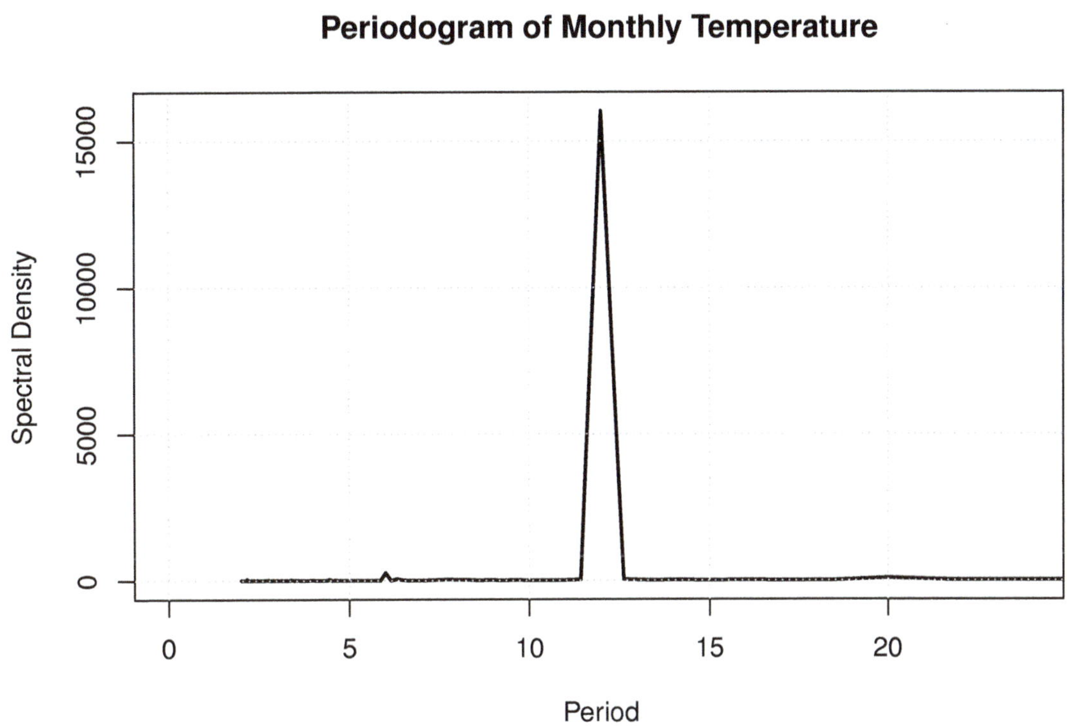

Figure 5.4: Periodogram of the monthly temperature series, showing a strong peak at a period of 12 months, indicating a clear seasonal cycle.

The periodogram (Figure 5.4) shows a strong peak at a period of 12 months, which confirms the presence of a clear seasonal cycle in the data. This means that the temperature tends to repeat every year, which is a common pattern in many environmental time series.

The main lesson is not the technical details behind the periodogram. The main lesson is that time series often repeat. If the repeating structure is strong, a simple forecast can sometimes work surprisingly well.

> ! Repeating Structure, Not Causation
>
> A strong peak in a periodogram does not imply causation. It indicates repeating structure. In this case, the annual cycle is driven by seasonal changes in solar radiation and climate.

5.6.2 A Simple Forecasting Example

The seasonal naive method is a very simple forecasting approach. If the data repeat every 12 months, we can use the previous year as a simple template for forecasting the next one. When past years show similar patterns, it is reasonable to expect the future to follow the same seasonal structure. This method is not complex, but it is honest, easy to explain, and often a useful starting point before trying more complex models.

> **! Before Forecasting**
>
> Ask what pattern you expect to continue. Should the next values be close to the most recent values, or should they resemble the same season from the previous year? A forecast should match the structure of the data.

```r
data(nottem) # load the nottem dataset
# use data up to 1935 for training
train_ts <- window(nottem, end = c(1935, 12))
# use data from 1936 onward for testing
test_ts <- window(nottem, start = c(1936, 1))

# Seasonal naive forecast: repeat the last 12 months
last_year <- tail(train_ts, 12)
# repeat the last year to match the length of the test set
fc_values <- rep(last_year, length.out = length(test_ts))
# convert to time series object
fc_ts <- ts(fc_values, start = start(test_ts), frequency = frequency(test_ts))
plot(fc_ts, # plot the forecast (thick blue line)
     type = "l",
     lwd = 3,
     col = "blue",
     main = "Seasonal Naive Forecast",
     xlab = "Time",
     ylab = "Temperature (°F)")
# add actual values for comparison
lines(test_ts, lwd = 1, col = "black") # observed values (thin black line)
grid() # add grid for better visualization
legend("topleft", bty = "n",
       legend = c("Forecast (thick)", "Observed (thin)"),
       col = c("blue", "black"), lwd = c(3, 1))
```

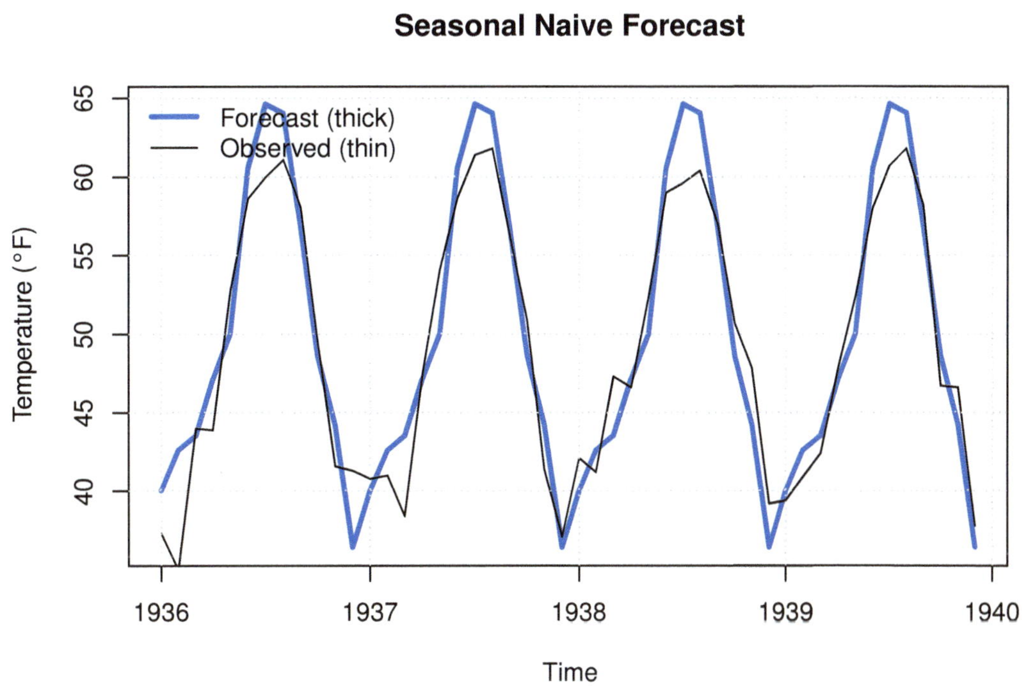

Figure 5.5: Seasonal naive forecast (thick blue line) compared with observed values (thin black line).

The forecast repeats the previous year's pattern, showing how seasonality can guide simple predictions (Figure 5.5). This is not a sophisticated forecast, but it is transparent and easy to interpret. That is often a good starting point. If the seasonal pattern is strong, even this simple method can perform reasonably well. More complex models may improve prediction accuracy, but they also require more assumptions and may be harder to interpret. So far, we used models that follow a fixed structure. Next, we explore models that learn rules directly from the data, rather than assuming a predefined form.

Researchers extend these forecasts with more advanced tools. Tsakiri, Marsellos, and Kapetanakis (2018) started from multiple linear regression on a river-discharge series and then added an artificial neural network, a flexible model you will meet beyond this book, which improved flood-forecast accuracy on the Mohawk River in New York.

The main idea is still the same: use existing patterns in data to help understand or estimate unknown values. Machine learning simply extends this idea using more flexible computational methods.

5.7 Introduction to Machine Learning

Machine learning methods are widely used in Earth and sustainability applications to identify patterns, improve prediction, and analyze complex environmental systems. Machine learning is another way to find patterns in data; it is not magic or artificial intelligence that "thinks" like a human. In simple terms, it is a group of methods that help computers recognize patterns in data.

Instead of being told the shape of the relationship in advance (like "fit a straight line"), the model figures out the pattern itself from the examples in the data. It can bend and adapt where a straight line cannot. Some machine learning methods can improve prediction accuracy, but they are often harder to interpret than simpler models.

A random forest is one example of this type of model. A decision tree is a flowchart of yes/no questions (such as "is it warmer than 80°F?") that ends in a prediction; a random forest builds many such trees and averages their results. This can improve prediction accuracy, but it also makes the model harder to interpret clearly than a simple regression model. In geoscience, this matters because we often need to explain not only what the model predicts, but also why we trust it.

Student research applies the same tools: Izaguirre Cortez et al. (2026) used multiple linear regression and a random forest in R to study which factors make a disaster rank as severe across a global disaster database, with Hurricane Sandy as a regional example.

Machine learning models do not automatically reveal causation. They detect patterns in the data. A strong predictive pattern may still be the result of shared seasonality, spatial structure, or a hidden variable. That is why critical thinking is essential.

5.7.1 A Simple Tree-Based Example

Machine learning models, like decision trees, learn patterns from data and use those patterns to make predictions.

The goal is the same as before: use known data to predict unknown values.

The difference is how each model represents the relationship:

- **linear regression** fits a straight line
- **decision trees** create if-then rules

> **!** Which Variables Matter Most?
>
> What rules might separate low ozone days from high ozone days? Would temperature, wind, or solar radiation matter more?

We will start with a simple decision tree so we can visualize how the model creates rules from the data.

```
# load the rpart package for decision tree modeling
library(rpart)
# fit a simple decision tree on the training data
tree_fit <- rpart(Ozone ~ Temp + Wind + Solar.R,
                  data = train,
                  # "anova" just tells rpart we are
                  # predicting a number, not a category
                  method = "anova")
```

Visualizing the Tree

```
# visualize the tree structure
# (the type/extra/fallen.leaves settings just control how the tree is drawn)
library(rpart.plot)
rpart.plot(tree_fit, type = 2, extra = 101,
           fallen.leaves = TRUE,
           main = "Decision Tree for Ozone Prediction")
# predict ozone on the test data
tree_pred <- predict(tree_fit, newdata = test)
# compare observed and predicted values: this is the tree's MAE
c(tree_MAE = mean(abs(test$Ozone - tree_pred)))
```

```
tree_MAE
17.59989
```

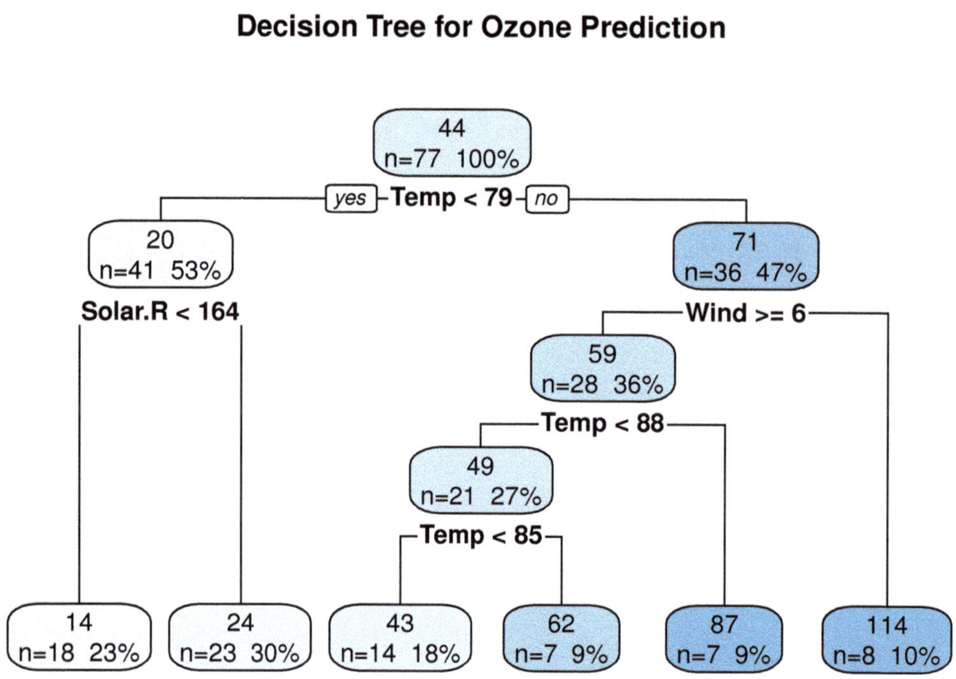

Figure 5.6: Decision tree model for ozone prediction. The tree splits the data into groups based on temperature, wind, and solar radiation, with each terminal node representing the average predicted ozone level for that set of conditions.

> How to read each box in the tree
>
> In each box, the top number is the predicted ozone, `n` is how many training days fall there, and the percentage is their share of the data. At a split, follow the left branch when the condition is true and the right branch when it is false.

Interpretation

- The tree (Figure 5.6) splits the data step by step into simpler groups based on rules (e.g., temperature → wind → solar radiation)
- Each branch represents one splitting rule (e.g., "Temp < 80°F" or "Wind > 10 mph")
- Each final node gives a **predicted ozone value**, and the two numbers below it are how many observations fall in that group (the count) and what percent of the data that is

The model learns these rules automatically by searching for splits that reduce prediction error in the training data. The tree first splits the data by temperature. If temperature is low, the model then examines wind. If temperature is high, it instead examines solar radiation. Each final branch gives a prediction for ozone. This makes the model relatively easy to interpret because each prediction follows a sequence of simple decision rules.

A tree-based model asks questions such as whether temperature is above or below a threshold, or whether wind is strong or weak. A random forest combines many trees, which usually makes predictions more stable. Even then, the model should still be checked against scientific reasoning.

The single number printed above is the **MAE** (mean absolute error), while the plot helps visualize *how* the model behaves. Together, they tell whether the tree is producing reasonable predictions or just a number without context.

5.7.2 Using the Tree for Prediction

The tree model can be used to predict ozone for new cases. The prediction is based on the rules learned from the training data. For example, if we have a new day with a certain temperature, wind, and solar radiation, we can follow the tree's rules to get a predicted ozone value.

Interpretation

- The model makes predictions for unseen data
- The average error is about **18 ppb**, slightly higher than regression

Observed vs Predicted

```
plot(test$Ozone, tree_pred,
     pch = 19,
     xlab = "Observed Ozone (ppb)",
     ylab = "Predicted Ozone (ppb)",
     main = "Tree Model: Observed vs Predicted")
abline(0, 1, lwd = 2)
grid()
```

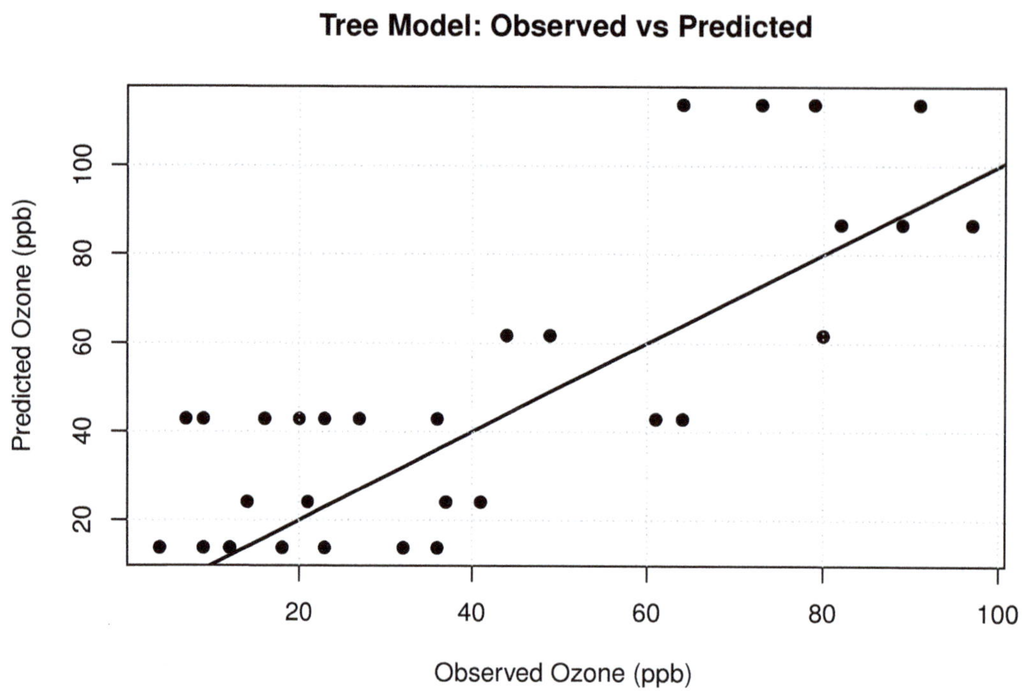

Figure 5.7: Observed versus predicted ozone values from the decision tree model.

The step-like prediction behavior is typical of tree-based models (Figure 5.7).

> **i** What do you notice?
>
> - Predictions form **horizontal bands**
> - The tree gives the **same value for groups of observations**
>
> This happens because the model uses rules instead of a continuous equation.

> **!** Key Insight
>
> - Linear regression → smooth relationship
> - Decision tree → step-like rules
>
> Both aim to predict unseen data, but in different ways.

⚠ Good predictions are not enough

A good prediction does not mean a correct explanation. Tree models may capture patterns without revealing true physical causes.

5.7.3 Tree-Based Models vs Linear Regression

Decision-tree models and linear regression can both predict ozone, but they represent relationships differently: one with a smooth line, the other with step-like rules.

Linear regression is like drawing one straight line through your data so it stays as close as possible to all the points overall.

Random forest regression extends the decision-tree idea by combining many trees and averaging their predictions. Each tree divides the data into groups using simple rules and produces its own prediction. The forest then averages the predictions from all trees. Because a single tree's prediction can change a lot if the data change slightly, averaging many trees steadies the result (reduces instability) and usually improves predictive performance. In this chapter, we begin with a single decision tree because its rules are easier to visualize and interpret.

- Linear regression → **one global straight line**
- Random forest → **many local decision rules combined together**

This flexibility can improve prediction accuracy, especially when relationships are more complicated than a simple straight line. However, it becomes more difficult to clearly trace how each prediction is made. The model may capture complex patterns, but it may not provide a clear explanation of the underlying processes.

! Simpler Models Are Often Better

In this book, we prioritize models that are **interpretable**. Understanding how a model works is often more valuable than small gains in accuracy.

Building intuition with simple models (like linear regression) provides the foundation needed to understand more flexible methods such as random forests. Similar machine learning workflows are widely used in sustainability analytics for climate prediction, flood risk assessment, environmental monitoring, and resource management.

Visual comparison: smooth vs step-like behavior

```
# Fit models on the same data for visualization
lm_fit <- lm(Ozone ~ Temp, data = train)

tree_fit_simple <- rpart(Ozone ~ Temp,
                         data = train,
                         method = "anova")

# Create ordered values for smooth plotting
ord <- order(train$Temp)
temp_sorted <- train$Temp[ord]

# Predictions
lm_pred <- predict(lm_fit,
                   newdata = data.frame(Temp = temp_sorted))
tree_pred <- predict(tree_fit_simple,
                     newdata = data.frame(Temp = temp_sorted))

# Plot
plot(train$Temp, train$Ozone,
     pch = 19,
     xlab = "Temperature (°F)",
     ylab = "Ozone (ppb)",
     main = "Linear vs Tree Model")

grid()

# Linear regression (smooth line)
lines(temp_sorted, lm_pred,
      lwd = 2)

# Tree model (step-like behavior)
lines(temp_sorted, tree_pred,
      lwd = 3, lty = 1, col = "darkgreen")
```

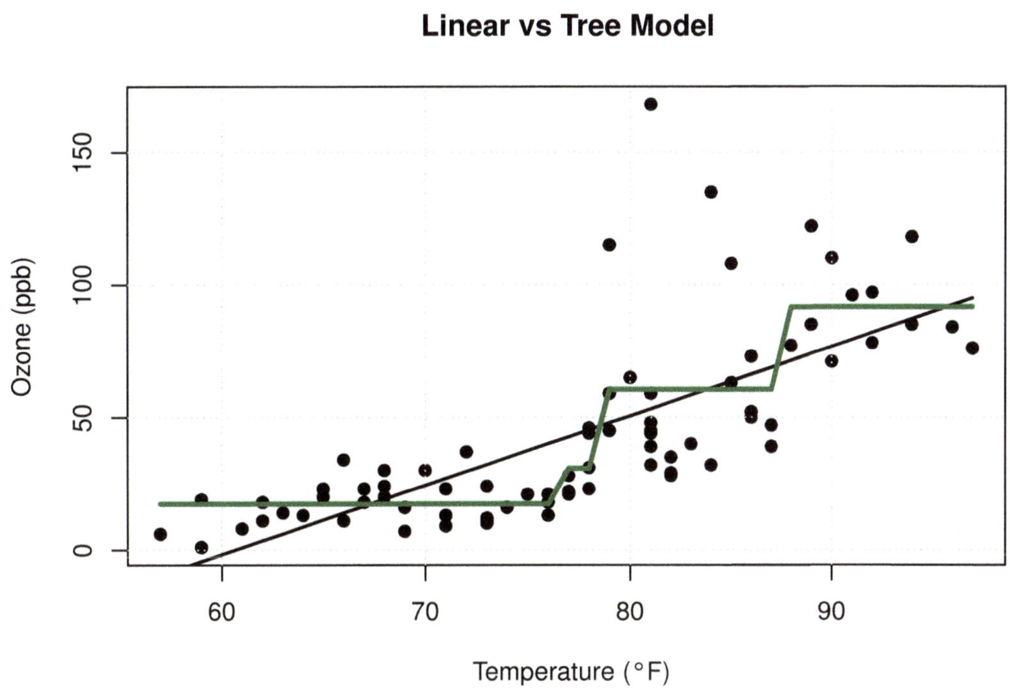

Figure 5.8: Comparison of linear regression (smooth line) and decision tree (step-like curve).

Smooth and rule-based prediction behaviors contrast clearly (Figure 5.8). Together with Figure 5.6 and Figure 5.7, this comparison highlights how different modeling approaches produce fundamentally different types of predictions. The straight line represents the linear regression (smooth and global), while the step-like curve represents the tree-based model (local and piecewise).

> **!** Accuracy Is Not Everything
>
> A more accurate model is not always a better scientific model. If we cannot explain it, we should be cautious about trusting it.

5.8 Complete Modeling Example

This example uses `airquality`, a classic environmental dataset. We will predict ozone using temperature, wind, and solar radiation. The purpose is not to build the most accurate possible model. The purpose is to show a complete and understandable workflow.

Before looking at the code, here is the scientific interpretation we expect to test:

- higher temperature may be associated with higher ozone
- stronger wind may reduce ozone by dispersing pollutants
- stronger solar radiation may be associated with higher ozone because sunlight supports photochemical reactions

Those are hypotheses, not guarantees. The model will help us check whether the data support them.

> **! Compare Expectations with Results**
>
> Write your expectation first. Then compare the model output with that expectation. If the result surprises you, do not ignore it. Ask whether the data, the model, or the interpretation needs to be reconsidered.

```
# Clean data
model_data <- na.omit(airquality[, c("Ozone",
                                     "Temp",
                                     "Wind",
                                     "Solar.R")])

# Visualize relationships first
par(mfrow = c(1, 3))
plot(model_data$Temp, model_data$Ozone,
     pch = 19,
     xlab = "Temperature (°F)",
     ylab = "Ozone (ppb)",
     main = "Ozone vs Temp")
grid()

plot(model_data$Wind, model_data$Ozone,
     pch = 19,
     xlab = "Wind (mph)",
     ylab = "Ozone (ppb)",
     main = "Ozone vs Wind")
grid()

plot(model_data$Solar.R, model_data$Ozone,
     pch = 19,
     xlab = "Solar Radiation",
     ylab = "Ozone (ppb)",
     main = "Ozone vs Solar Radiation")
```

```
grid()
```

```
par(mfrow = c(1, 1))
```

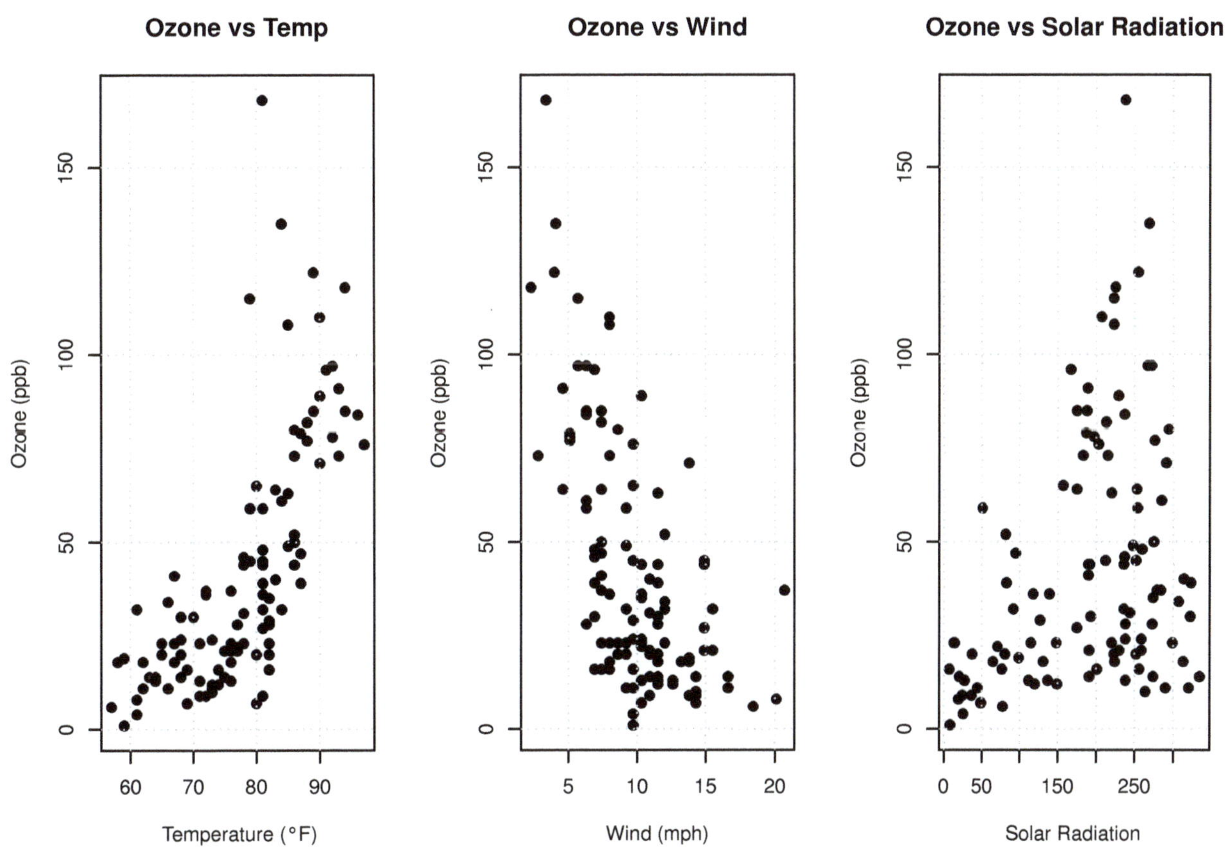

Figure 5.9: Ozone plotted against temperature, wind speed, and solar radiation.

These plots help us inspect the relationships before modeling. They also help us see whether the relationships appear clear, weak, or noisy. A model should never come before understanding the data visually.

To evaluate the model fairly, we again separate the data into training and testing sets (as defined earlier).

```
set.seed(123)
n <- nrow(model_data)
train_id <- sample(seq_len(n), size = floor(0.7 * n))
train <- model_data[train_id, ]
test <- model_data[-train_id, ]
```

```r
model_fit <- lm(Ozone ~ Temp + Wind + Solar.R, data = train)
summary(model_fit)
```

```

Call:
lm(formula = Ozone ~ Temp + Wind + Solar.R, data = train)

Residuals:
    Min      1Q  Median      3Q     Max 
-29.912 -16.086  -4.016   7.824  92.987 

Coefficients:
             Estimate Std. Error t value Pr(>|t|)    
(Intercept) -74.18853   28.90334  -2.567   0.0123 *  
Temp          1.84998    0.32151   5.754 1.91e-07 ***
Wind         -3.34508    0.80999  -4.130 9.55e-05 ***
Solar.R       0.04507    0.02867   1.572   0.1202    
---
Signif. codes:  0 '***' 0.001 '**' 0.01 '*' 0.05 '.' 0.1 ' ' 1

Residual standard error: 22.4 on 73 degrees of freedom
Multiple R-squared:  0.6131,	Adjusted R-squared:  0.5972 
F-statistic: 38.56 on 3 and 73 DF,  p-value: 4.82e-15
```

If the coefficient for temperature is positive, the model says ozone tends to increase with temperature after accounting for wind and solar radiation. If wind has a negative coefficient, the model says stronger wind tends to be associated with lower ozone. That is useful, but it still does not prove causation.

```r
# Test-set evaluation
model_pred <- predict(model_fit, newdata = test)

mae_model <- mean(abs(test$Ozone - model_pred))
rmse_model <- sqrt(mean((test$Ozone - model_pred)^2))

mae_model
```

```
[1] 15.53134
```

```r
rmse_model
```

```
[1] 19.27173
```

```
# Observed vs predicted values
plot(test$Ozone, model_pred,
     pch = 19,
     xlab = "Observed Ozone (ppb)",
     ylab = "Predicted Ozone (ppb)",
     main = "Test Set: Observed vs Predicted")
abline(0, 1, lwd = 2)
grid()
```

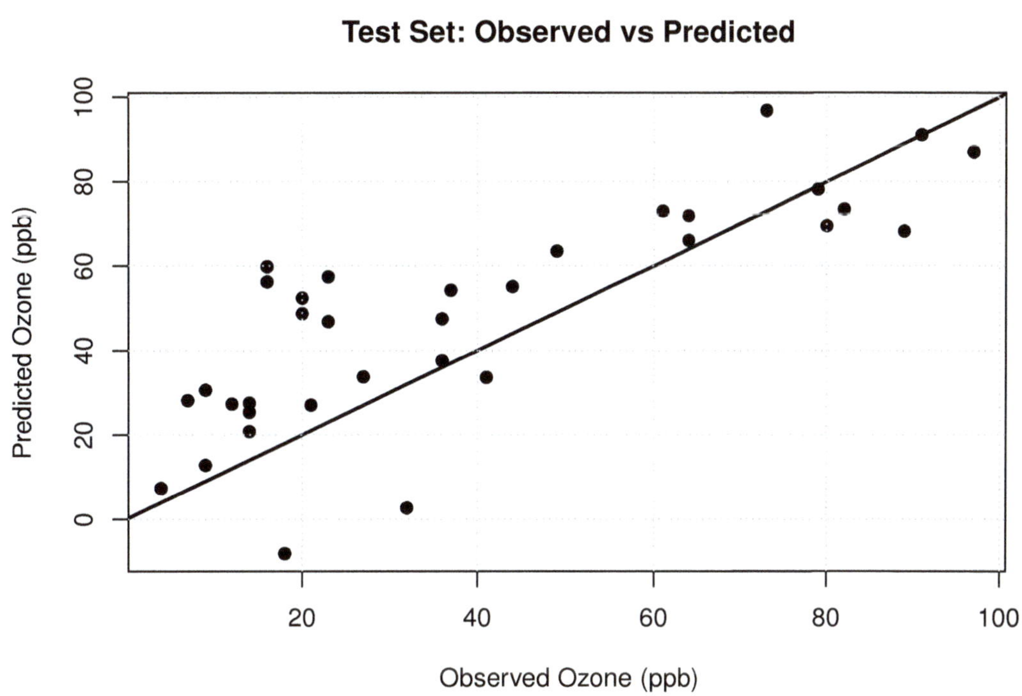

Figure 5.10: Observed versus predicted ozone values for the final regression model.

The observed and predicted ozone values on the test dataset can be compared directly (Figure 5.10). If the points cluster near the 1:1 line, the model is doing a reasonable job. If they are widely scattered, the model is missing important information and has limited predictive power. The result should be judged in scientific context, not only by one number.

5.8.1 Interpretation of Results

The results should be read as a scientific summary, not just as output from software. The model results may indicate that a relationship appears positive or negative, strong or weak, stable or unstable. It can also show which variables matter more in the sample we studied.

At the same time, there are limits:

- the model may not generalize to a different region
- the model may miss nonlinear behavior
- missing variables may distort the result
- correlations may reflect a shared driver rather than a direct effect
- a good prediction does not automatically mean a correct explanation

That is why interpretation must stay close to the data and the science.

5.9 Common Mistakes to Avoid

- **Overfitting**: using a model that is too complex for the amount of data
- **Misinterpreting relationships**: reading correlation as causation
- **Blind trust in models**: assuming the computer is always right
- **Ignoring evaluation**: checking fit on the training data only
- **Forgetting context**: separating the model from the geoscience problem

A useful model is not only accurate. It is also understandable, defensible, and scientifically reasonable.

> **ℹ No Model Is Perfect**
>
> Even useful models contain uncertainty, assumptions, and limitations. The goal is not perfect prediction, but better understanding.

5.10 What Models Can and Cannot Do

Models are simplified representations of reality. They are useful because they help us explain patterns and make predictions, but they are never perfect.

The main ideas in this chapter are:

- start with visualization
- keep models simple when possible
- interpret coefficients in plain language
- evaluate on new data
- respect time order in time series
- treat machine learning as pattern learning, not truth
- remember that correlation is not causation

Good modeling is not only about accurate prediction. It is also about understanding what the model is saying, what it may be missing, and whether the result makes scientific sense.

> **i** Models Have Limitations
>
> In geoscience, the best model is often the one that is clear, useful, and honest about its limitations.

Up to this point, we have learned individual analytical tools. The next chapter demonstrates how these tools can be combined to investigate real Earth-system problems involving tectonics, flooding, and climate change.

> Where These Methods Lead
>
> In this chapter you built models that predict one measurement from others, and you saw how a simple model can be tested and improved. Researchers use these same steps on real environmental problems.
>
> On the Mohawk River in New York, Tsakiri, Marsellos, and Kapetanakis (2018) began with the same kind of regression you fitted here, using past measurements to predict the river's flow. They then added a more flexible tool, an artificial neural network, that learns patterns a straight-line model alone can miss. Together the two models predicted the river's rise more accurately than regression on its own, which means earlier and clearer warning of when the river is likely to flood. Better warning gives the towns along the river more time to prepare and protect people and property.
>
> A related student study makes a different point. Izaguirre Cortez and colleagues (2026) compared regression with a random forest, a model built from many small decision trees like the one you grew here, on a global record of past disasters, and found that which events a model flags as most "severe" depends partly on which measurements it is given. Hurricane Sandy always counted among the worst disasters, but its exact ranking shifted with those choices, a reminder that a model's design shapes its answer.
>
> The path is the same one you have just walked: start simple, check how well the model really works, and use a more flexible tool only when a simpler one is not enough.

5.11 From Analysis to Poster

When presenting model results on a poster or slide, focus on clarity:

- show the question first
- include one clean plot of the key relationship
- report the main model result in one sentence
- explain what the model suggests in plain language
- state at least one limitation
- avoid overwhelming the audience with too many formulas or diagnostics

A good poster tells a story: problem, model, result, meaning, and caution.

5.12 Questions to Consider Before Continuing

Before moving to the exercises, ask yourself:

- What is the model simplifying?
- What might the model be missing?
- Would I trust this prediction outside the range of the observed data?
- Is the model helping me understand the system, or only producing a prediction?
- Which assumptions of the model could affect my conclusions?

5.13 Exercises

1. Use `airquality` to fit a simple regression model with one predictor only. Compare the result with the multivariable model.
2. Create a train/test split with a different random seed and see whether the error changes.
3. Fit a model to a different environmental variable in `airquality` or another built-in dataset and interpret the coefficients in plain language.
4. Use the monthly temperature series and try a different simple forecast, such as using the previous month only. Compare it with the seasonal naive forecast.
5. Write two sentences explaining why a strong correlation does not prove causation in a geoscience study.

5.13.1 References (Data and Further Reading)

1. James, G., Witten, D., Hastie, T., & Tibshirani, R. *An Introduction to Statistical Learning.* https://www.statlearning.com

2. Montgomery, D. C., Peck, E. A., & Vining, G. G. *Introduction to Linear Regression Analysis.*

3. Hyndman, R. J., & Athanasopoulos, G. *Forecasting: Principles and Practice.* https://otexts.com/fpp3

4. rpart package documentation. https://cran.r-project.org/package=rpart

5. rpart.plot package documentation. https://cran.r-project.org/package=rpart.plot

6. Tsakiri, K., Marsellos, A., & Kapetanakis, S. (2018). *Artificial Neural Network and Multiple Linear Regression for Flood Prediction in Mohawk River, New York.* Water, 10(9), 1158. https://doi.org/10.3390/w10091158

7. Izaguirre Cortez, B., Rezki, A., Najarro Lima, M., & Marsellos, A. E. (2026). *A.I., Machine Learning, and Ethical Decision-Making in Sustainable Infrastructure: A Hurricane Sandy Modeling Perspective.* In 33rd Annual Conference on the Geology of Long Island and Metropolitan New York, April 2026.

8. Plitnick, T. A., Marsellos, A. E., & Tsakiri, K. G. (2018). *Time Series Regression for Forecasting Flood Events in Schenectady, New York.* Geosciences, 8(9), 317. https://doi.org/10.3390/geosciences8090317

9. Al-Saeedi, K., Zhou, D., Fish, A., Tsakiri, K., & Marsellos, A. (2025). *A Methodological Comparison of Forecasting Models Using KZ Decomposition and Walk-Forward Validation.* Mathematics, 13(21), 3410. https://doi.org/10.3390/math13213410

10. Al-Saeedi, K., Fish, A., Zhou, D., Tsakiri, K., & Marsellos, A. (2026). *Multi-Scale Decomposition and Autocorrelation Modeling for Classical and Machine Learning-Based Time Series Forecasting.* Mathematics, 14(2), 283. https://doi.org/10.3390/math14020283

Part III

Applications

Chapter 6

Earth System Case Studies

Applying Data Analysis to Real-World Geoscience Problems

6.1 Introduction

In the previous chapters, we learned how to work with spatial data, time series, and analytical workflows. In this chapter, we combine those ideas to study Earth systems at much larger spatial and temporal scales.

Many sustainability challenges are not isolated to one location or one moment in time. Climate change, greenhouse gas emissions, sea-level rise, and environmental change operate across regions, continents, and decades.

The goal of this chapter is to understand how changes through space and time work together in Earth systems. So far, we have explored individual tools: data handling, visualization, spatial analysis, time series, and modeling. In real-world geoscience problems, however, these tools are rarely used in isolation. Instead, they are combined into a workflow that moves from raw data to meaningful interpretation.

Working with real data introduces new challenges. Data may be incomplete, noisy, or influenced by multiple processes at once. Real-world data are often messy. Different methods may emphasize different aspects of the same dataset. Because of this, the goal is not to apply as many techniques as possible, but to choose appropriate methods and understand what the results mean.

Each case study in this chapter follows a consistent structure: - defining the problem, - preparing the data, - selecting appropriate methods, - analyzing results, - and interpreting findings in a scientific context.

The emphasis is on reasoning and interpretation, not technical complexity. Simple methods, when used carefully, are often more powerful than complex ones that are not fully understood.

Unlike the previous chapters, this chapter is organized around complete real-world applications rather than beginner, intermediate, and advanced sections. Each case study combines multiple concepts into a single analytical workflow. Although the datasets differ, all three case studies address the same underlying question:

How can we use data to understand change in Earth systems through space and time?

The GPS example focuses on tectonic deformation, the flooding example examines coastal vulnerability, and the Greenland example investigates climate-driven ice loss. Together, they illustrate how spatial and temporal evidence can be combined to understand complex environmental processes. They also highlight an important sustainability challenge: environmental change affects both natural systems and human communities. Understanding where change occurs, how it evolves through time, and what consequences may follow is essential for informed planning, hazard mitigation, resource management, and long-term environmental decision-making.

> **i What you will be able to do**
>
> By the end of this chapter, you will be able to:
>
> - clean a real GPS time series: fill missing dates, flag outliers with Z-scores (how far a value sits from the average), and interpolate the gaps
> - examine a series with autocorrelation, a periodogram, a moving average, and cross-correlation to find memory, cycles, trends, and shared movement
> - split a signal into trend, seasonal cycle, and remainder, then remove the trend or the season to isolate the part you want to study
> - estimate flood exposure from a digital elevation model (a map of ground height at every point) by counting flooded land at several water levels and mapping it over a hillshade
> - open a NetCDF satellite file and turn its spatial grid into a single time series by averaging the cells inside a region
> - compare two dates with a difference map and a box plot to see where change happens and how it repeats by month

6.2 Case Study 1: Ground Displacement and Tectonic Motion

GPS stations continuously measure how the Earth's surface moves through time. Even tiny ground movements can help scientists study how tectonic plates move and deform the Earth's crust. In this case study, we analyze GPS displacement data to investigate how the ground changes through time and how different analytical tools help us interpret those patterns.

This mirrors recent student research. Moore et al. (2021) used R to analyze GPS records from long-running stations on Long Island, New York, and measure a very slow, long-term sinking of the land.

6.2.1 Loading and Cleaning the GPS Data

We use GPS station time series data from the Nevada Geodetic Laboratory at the University of Nevada, Reno (http://geodesy.unr.edu), which publishes daily position records for thousands of stations worldwide. Each record gives how far the station has moved east, north, and up (in meters) compared to a fixed starting position. We focus on a single station near New York City (NYCI).

```r
library(zoo)

# Download GPS data directly from the Nevada Geodetic Laboratory
url <- "https://geodesy.unr.edu/gps_timeseries/IGS20/tenv3/IGS20/NYCI.tenv3"
gps_raw <- read.table(url, header = TRUE)

# Build a simple data frame with just date, east, north, and up.
# The original column names contain symbols, so R automatically renamed them
# (for example, _e0(m) became X_e0.m.). We use the R-friendly names here.
gps <- data.frame(
  date  = as.Date(gps_raw$X__MJD, origin = "1858-11-17"),
  east  = gps_raw$X_e0.m.    + gps_raw$X__east.m.,
  north = gps_raw$X____n0.m. + gps_raw$X_north.m.,
  up    = gps_raw$u0.m.      + gps_raw$X____up.m.
)

head(gps)
```

```
        date     east   north        up
1 2006-04-25 192.3668 4513991 -15.12136
2 2006-04-26 192.3668 4513991 -15.12021
3 2006-04-27 192.3671 4513991 -15.12603
4 2006-04-28 192.3664 4513991 -15.12891
```

```
5 2006-04-29 192.3669 4513991 -15.12864
6 2006-05-03 192.3681 4513991 -15.12018
```

The download takes about five seconds. The unusual column names like `_e0(m)` and `__east(m)` are how the Nevada Geodetic Laboratory stores GPS positions: each coordinate is split into a reference value (`_e0`) and a small offset (`__east`). Adding them together gives the actual position in meters.

> **i About the date column**
>
> The `__MJD` column stores dates as Modified Julian Day numbers, counted from November 17, 1858. The `as.Date()` function converts these to ordinary calendar dates when we tell it the correct origin.

A small number of NAs may remain at the very start or end of the series if those edge values were flagged as outliers. `na.approx()` interpolates (fills gaps between known points) but does not extrapolate (estimate values beyond the first or last observation). In the cleaning step below we simply trim those few edge points, so the series we analyze is complete and gap-free.

Real GPS data contain missing days and occasional outliers caused by satellite errors, equipment problems, or atmospheric interference. We will clean the data using techniques introduced in Chapter 2: filling missing dates, removing outliers with Z-scores (a measure of how far a value sits from the average), and interpolating the gaps.

```r
# Step 1: Fill in missing dates so the time series is continuous
all_dates <- seq(min(gps$date), max(gps$date), by = "day")
gps_full <- merge(data.frame(date = all_dates), gps, by = "date", all.x = TRUE)

# Step 2: Mark extreme values (more than 3 standard deviations from the
# mean) as NA
z_score <- function(x) (x - mean(x, na.rm = TRUE)) / sd(x, na.rm = TRUE)

gps_full$east [abs(z_score(gps_full$east))  > 3] <- NA
gps_full$north[abs(z_score(gps_full$north)) > 3] <- NA
gps_full$up   [abs(z_score(gps_full$up))    > 3] <- NA

# Step 3: Interpolate missing values using the zoo package (introduced in
# Chapter 2)
gps_cleaned <- gps_full
gps_cleaned$east  <- na.approx(gps_cleaned$east,  na.rm = FALSE)
gps_cleaned$north <- na.approx(gps_cleaned$north, na.rm = FALSE)
gps_cleaned$up    <- na.approx(gps_cleaned$up,    na.rm = FALSE)
```

```
# Step 4: Interpolation cannot fill gaps at the very start or end of the
# series, so a few NAs may still sit at the edges. We trim those few rows
# so the series is complete and ready for every analysis step that follows.
gps_cleaned <- na.omit(gps_cleaned)

head(gps_cleaned)
```

```
        date     east   north        up
1 2006-04-25 192.3668 4513991 -15.12136
2 2006-04-26 192.3668 4513991 -15.12021
3 2006-04-27 192.3671 4513991 -15.12603
4 2006-04-28 192.3664 4513991 -15.12891
5 2006-04-29 192.3669 4513991 -15.12864
6 2006-04-30 192.3672 4513991 -15.12652
```

> When column names look strange
>
> The Nevada Geodetic Laboratory file uses column headers with symbols and underscores that are not valid R variable names. read.table() rewrites these automatically. For example, `__MJD` becomes `X__MJD`, while `_e0(m)` is converted to `X_e0.m.`; both names are safe to use in R. You can always run `names(gps_raw)` to see what R has called each column.

We now have a clean, continuous time series of east, north, and up displacements ready for analysis.

6.2.2 Why cleaning matters

To recap why the steps above were necessary: the raw GPS data contain missing values and outliers, which can distort our analysis. We first identify and remove outliers using a Z-score method, replacing them with NA. Then, we interpolate missing values to create a continuous time series. Cleaning helps reduce misleading patterns caused by missing values, noise, or sudden spikes due to instrument or environmental effects, and clarifies the underlying patterns in the data.

> Think before removing
>
> When removing outliers, do not break the continuity of the time series. Simply deleting values can distort the temporal structure and affect downstream analysis. Instead, replace removed values with `NA` and use appropriate interpolation methods to preserve continuity while minimizing bias.

Why Missing Values Matter

For example, a missing-values dataset is like recording a daily process but skipping a few days each week. The timeline becomes compressed, and a pattern that originally repeats every 10 days may appear to repeat every 7 days. This can create the false appearance of a weekly cycle that does not reflect the true behavior of the system.

Interpolation Is an Estimate

Interpolation estimates missing values. It does not recover the true measurements.

```
par(mfrow=c(1,2))
plot(gps$date, gps$up, type = "l",
     col = "blue",lwd=0.3,
     main = paste("Raw: GPS - Up"),
     xlab = "Date", ylab = "Up")

plot(gps_cleaned$date, gps_cleaned$up, type = "l",
     col = "blue",lwd=0.3,
     main = paste("Cleaned & Interpolated: GPS - Up"),
     xlab = "Date", ylab = "Up")
```

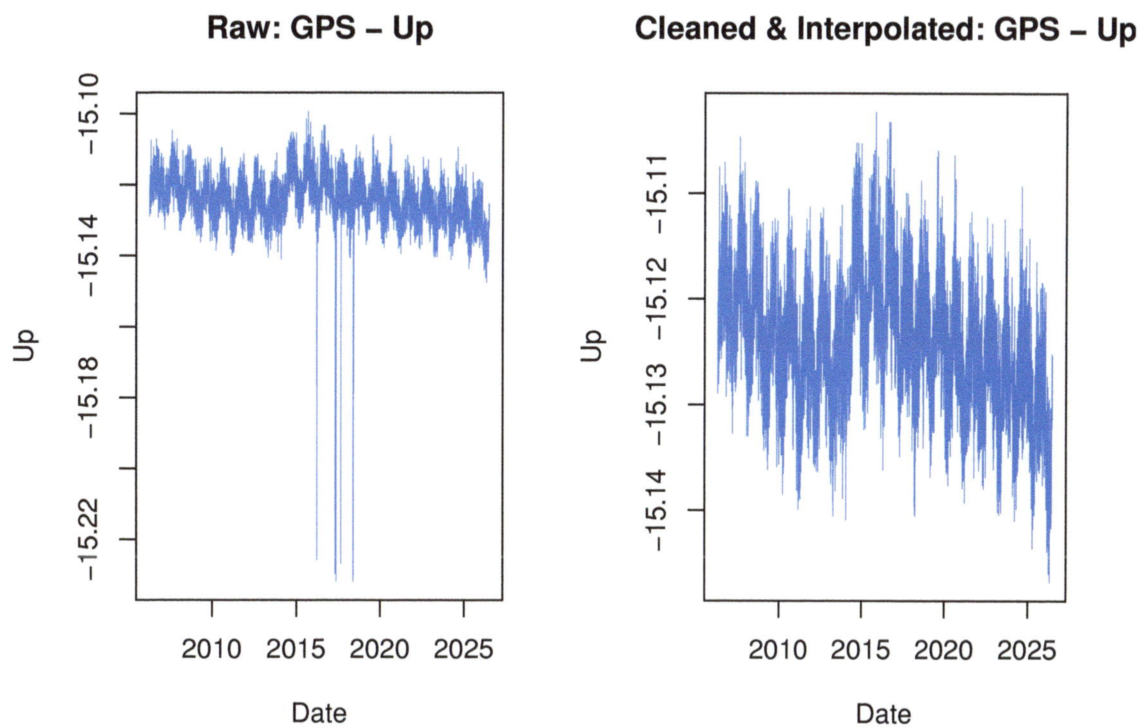

Figure 6.1: Comparison of raw (left) and cleaned (right) GPS vertical displacement (Up component).

The raw signals (left) and the cleaned signals (right) can be compared directly (Figure 6.1). The **raw GPS** signal contains irregular spikes (outliers) and discontinuities, which are likely due to measurement errors or missing observations. After cleaning and interpolation, the **cleaned & interpolated** time series becomes more consistent, allowing underlying patterns to be more clearly observed. This step is essential, as unprocessed data can lead to misleading conclusions.

> **i** Check Your Understanding
>
> Before continuing, make sure you understand:
>
> - why missing values can distort a time series,
> - why outliers are replaced with `NA` instead of deleted,
> - and why interpolation estimates values rather than recovering true measurements.

6.3 Exploring Patterns in the Signal

Now that the data are clean, we ask a series of simple questions about them: Does the past predict the present? Are there repeating cycles? What is the slow, long-term trend? Each tool below answers one of these questions.

6.3.1 Autocorrelation

In many Earth systems, current conditions are often influenced by previous conditions. Autocorrelation simply checks how strongly today's value resembles the values from previous days. It helps us understand whether the system has memory, that is, whether current displacement is influenced by earlier observations. In geophysical processes, this is often expected because changes in the Earth's surface occur gradually rather than abruptly. By examining how values relate to earlier observations, we can identify persistence and repeating structure in the time series.

> ⚠ Think before interpreting
>
> A slow decay in autocorrelation can be confusing, making it unclear whether to focus on lag 1, 2, or beyond (where lag refers to a time delay between observations). This gradual decline does not imply a specific physical cause. It indicates dependence, but that dependence may arise from multiple processes, such as trend or seasonality. Always consider what underlying mechanism could produce the observed pattern.

> System Memory
>
> Imagine pushing a heavy object. Once it starts moving, it does not stop immediately; it continues due to momentum. In a similar way, today's value still reflects the influence of previous days, creating a gradual, connected behavior over time.

```
acf(gps_cleaned$up, main = "ACF of Cleaned GPS Up")
```

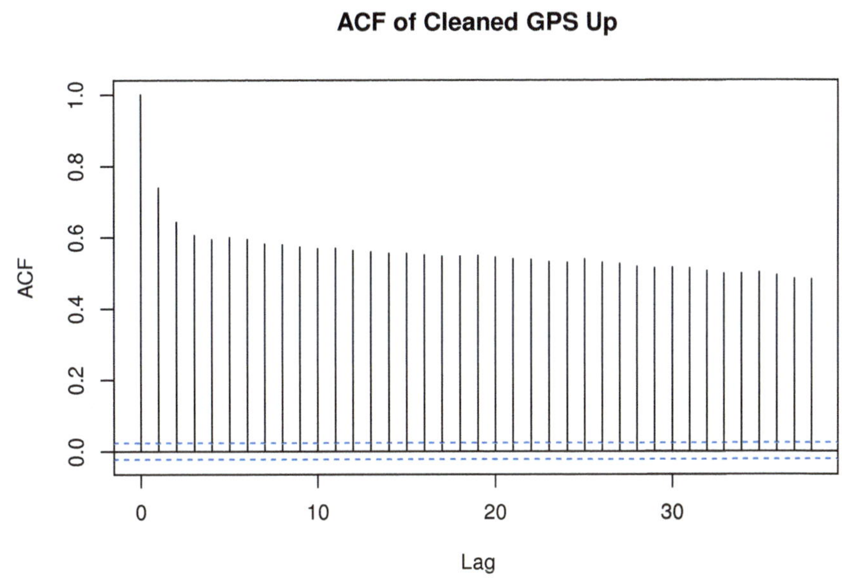

Figure 6.2: Autocorrelation function of the cleaned GPS vertical displacement showing strong temporal dependence.

The **autocorrelation** function (acf) shows a slow decay (Figure 6.2), indicating strong dependence between observations over time. Today's value is strongly related to earlier values, including observations from several previous days. As the lag increases, the dependence gradually decreases through time. This suggests that the motion is not random but changes gradually through time, showing persistence rather than random noise.

> **!** What Autocorrelation Does Not Tell Us
>
> Autocorrelation does not explain *why* the system behaves this way. It only tells us that nearby observations are related through time.

6.3.2 Spectral Analysis

Spectral analysis looks for repeating cycles by measuring which rhythms (for example, a once-a-year pattern) are strongest in the data. In other words, it sorts the small ups and downs in the signal by how often they repeat. While patterns may be visible in a time plot, this method helps confirm whether those patterns are truly periodic and reveals their dominant time scales.

> ⚠ Think before concluding
>
> A strong peak in the periodogram does not necessarily indicate a tectonic signal. Many periodic patterns arise from environmental influences, such as seasonal temperature changes, hydrological loading, or ionospheric fluctuations. Always distinguish between physical processes and recurring external effects.

> Finding Repeating Cycles
>
> Imagine a heartbeat. The regular beating creates a strong periodic signal at a specific frequency. However, if you were to analyze the sound of the heartbeat, you might also detect other periodic noises, such as breathing or background sounds. Similarly, in GPS data, we may find strong periodic signals that are not related to tectonic motion but rather to environmental factors that repeat regularly.

A strong peak means that a repeating cycle is very important in the data. In our case, we find a strong peak near a period of one year, which suggests that there is a dominant annual cycle in the GPS displacement.

```
# periodogram() finds the strongest repeating cycles;
# below we plot the period (1/freq)
sp <- TSA::periodogram(gps_cleaned$up, plot = F)
plot(1/sp$freq,sp$spec, type = "l",
     xlab = "Period (days)",
     ylab = "Spectral Density",
     main = "Periodogram of Cleaned GPS Up",xlim=c(0,500))
# Mark one year (365 days): the tall peak sits right next to this line
abline(v = 365, lty = 2, col = "red")
text(365, max(sp$spec, na.rm = TRUE), "1 year (365 days)", pos = 4, col = "red")
```

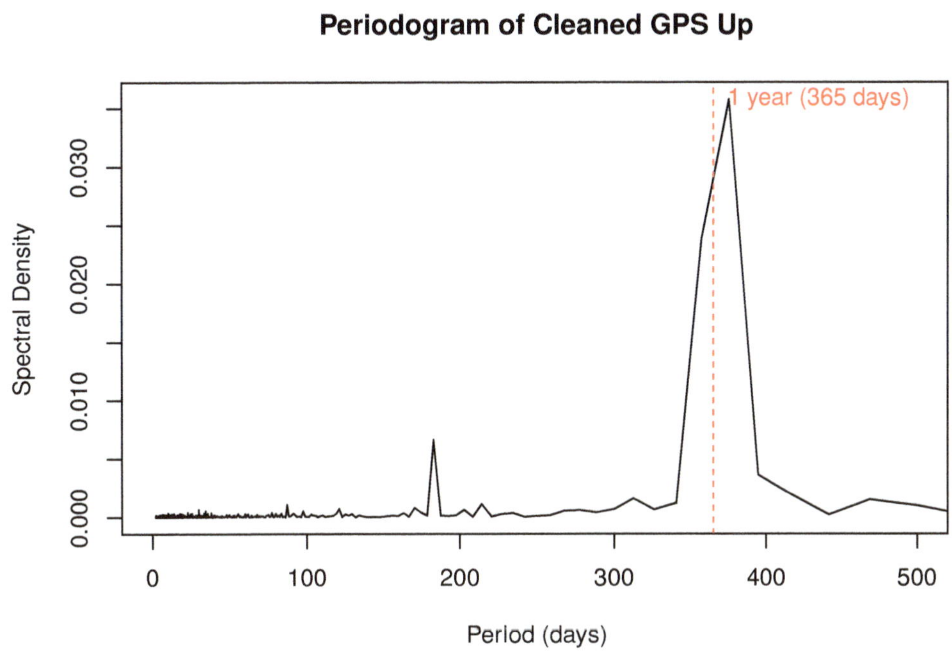

Figure 6.3: Periodogram of the cleaned GPS vertical displacement highlighting a dominant annual cycle.

The **periodogram** (Figure 6.3) reveals a strong peak near a period of one year, indicating a dominant annual cycle in the displacement signal. This is commonly associated with seasonal effects such as temperature variations, hydrological loading, or atmospheric influences, rather than tectonic motion alone. Identifying these periodicities can help differentiate between tectonic signals and environmental noise.

6.3.3 Moving Average

A moving average smooths a time series by reducing short-term fluctuations. It helps reveal slower long-term behavior while reducing rapid short-term variability. This makes it easier to identify gradual trends, seasonal patterns, or multi-year changes that may otherwise be hidden within noise. However, smoothing does not only remove "noise"; it can also remove real physical signals that occur on shorter timescales.

In signal processing, this type of smoothing is often called a low-pass filter, because it lets slow, long-term changes pass through while removing fast, short-term ones.

For example, a large window such as 365 days can strongly weaken or remove:

- Lunar and tidal cycles
- Weather and atmospheric variability
- Hydrological signals such as rainfall, runoff, and soil moisture changes
- Seismic swarms and short-term tectonic or volcanic activity

- Anthropogenic influences such as reservoir loading or fluid injection or hydraulic fracturing (fracking)

⚠ Smoothing Can Hide Real Events

Too much smoothing can remove meaningful information, not just noise. A large smoothing window can cause short-term events to disappear, blur transitions, weaken intermediate cycles, and distort timing information. In particular, some moving averages (such as trailing windows) can shift features in time, making events appear delayed. Even when no shift is introduced, smoothing can spread a sharp event over a wider period, making it unclear exactly when it occurred. Always choose the smoothing window based on the timescale of the process you want to study. Too little smoothing may leave noise, while too much smoothing may hide important variability. In geophysical and environmental systems, many important processes are episodic, transient, or multi-scale. Applying a large moving average may unintentionally eliminate the very signals you are trying to study.

What Smoothing Does

Imagine observing a landscape through a blurred lens. A slight blur removes small irregularities, and a lake may appear as a valley. However, a stronger blur may make the entire landscape look flat, erasing hills, valleys, and important features. Similarly, a moving average can clarify long-term trends, but if applied too aggressively, it can erase meaningful structures.

```
# rollmean() comes from the zoo package (introduced in Chapter 2)
# If you get an error, run: library(zoo) to load the library

plot(gps_cleaned$date, rollmean(gps_cleaned$up, k = 365, fill = NA),
     type = "l", col = "red",
     main = "Rolling Mean (12 months) of GPS Up",
     xlab = "Date", ylab = "Up (Rolling Mean)")
```

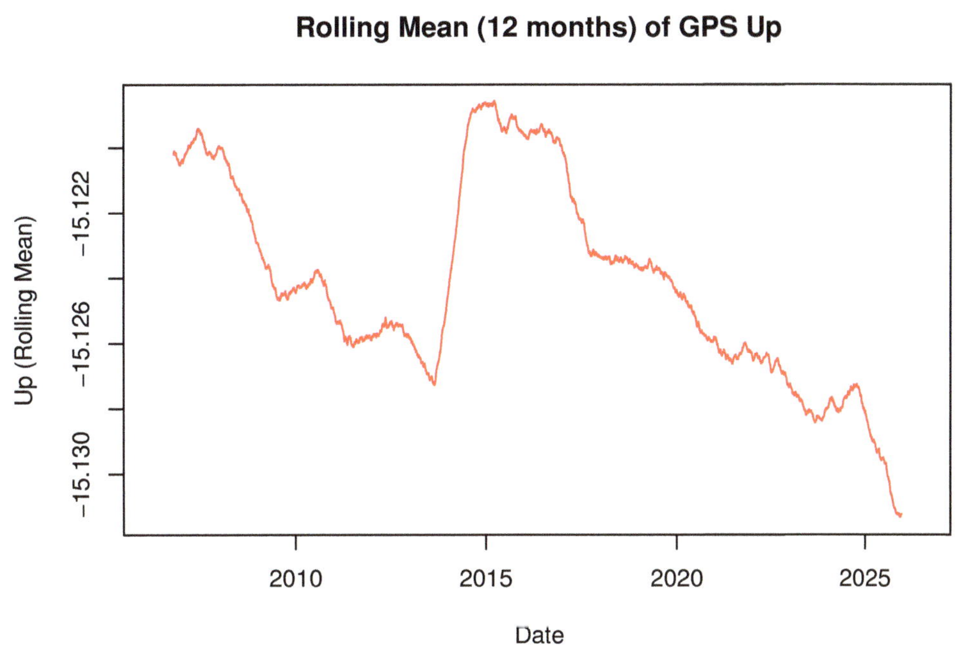

Figure 6.4: Rolling mean (365-day window) of GPS vertical displacement highlighting long-term trends.

The **rolling mean** (Figure 6.4) removes short-term variability and highlights long-term trends in displacement. This trend may reflect slow long-term deformation of the Earth's surface. However, the strong smoothing also eliminates short-term events, such as seismic swarms, volcanic activity, or rapid hydrological changes, which may be critical for understanding underlying processes. The smoothed signal emphasizes long-term structure but may no longer reflect important short-term dynamics.

For example, a sudden GPS displacement caused by an earthquake may appear smoother and more gradual after strong smoothing is applied. Short-lived events, such as seismic swarms or rapid hydrological changes, may also become less visible, appear spread over a longer period, or show reduced apparent magnitude after smoothing.

> **i Connecting the Methods**
>
> So far, we have explored:
>
> - autocorrelation → does the system have memory?
> - spectral analysis → are there repeating cycles?
> - moving average → what long-term behavior becomes visible after smoothing?
>
> These methods look at different aspects of the same signal.

6.3.4 Cross-Correlation

Cross-correlation compares two time series to see if they move together, or whether one tends to follow the other after some lag. In our case, it helps us explore how vertical and horizontal displacements interact, and whether they occur simultaneously or one responds after the other.

> ⚠ Think before assuming cause
>
> A strong cross-correlation does not necessarily indicate a direct causal relationship. Both series may be influenced by a common underlying process, such as tectonic deformation, which causes them to move together without one directly causing the other. Always consider the possibility of shared drivers or external factors that could produce correlated behavior.

> Shared Causes Can Create Correlation
>
> Imagine two dancers moving in sync. They may not be leading or following each other, but instead both responding to the same music. Similarly, vertical and horizontal displacements may be correlated because they are both responding to the same tectonic forces, rather than one directly causing the other.

```r
ccf(gps_cleaned$up, gps_cleaned$east, main = "CCF: GPS Up vs East")
```

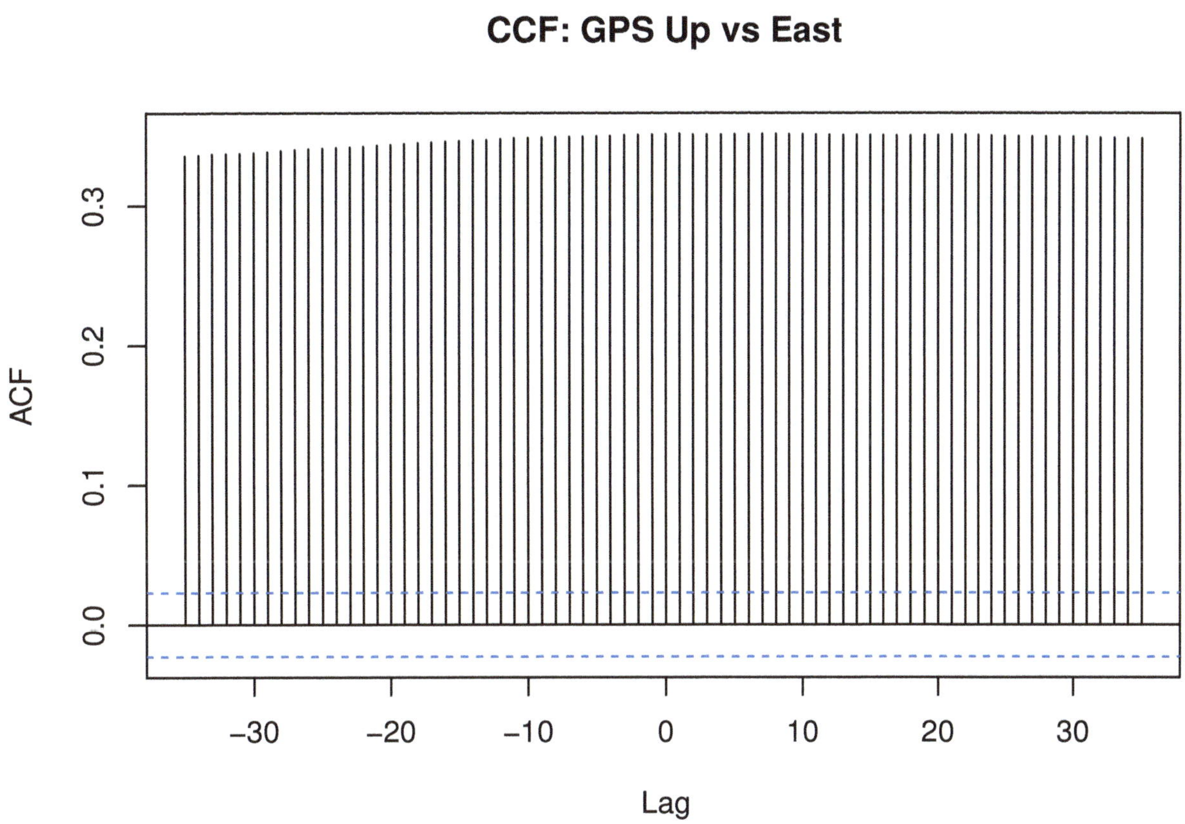

Figure 6.5: Cross-correlation between vertical (Up) and eastward displacement.

```r
ccf(gps_cleaned$up, gps_cleaned$north,
    main = "CCF: GPS Up vs North")
```

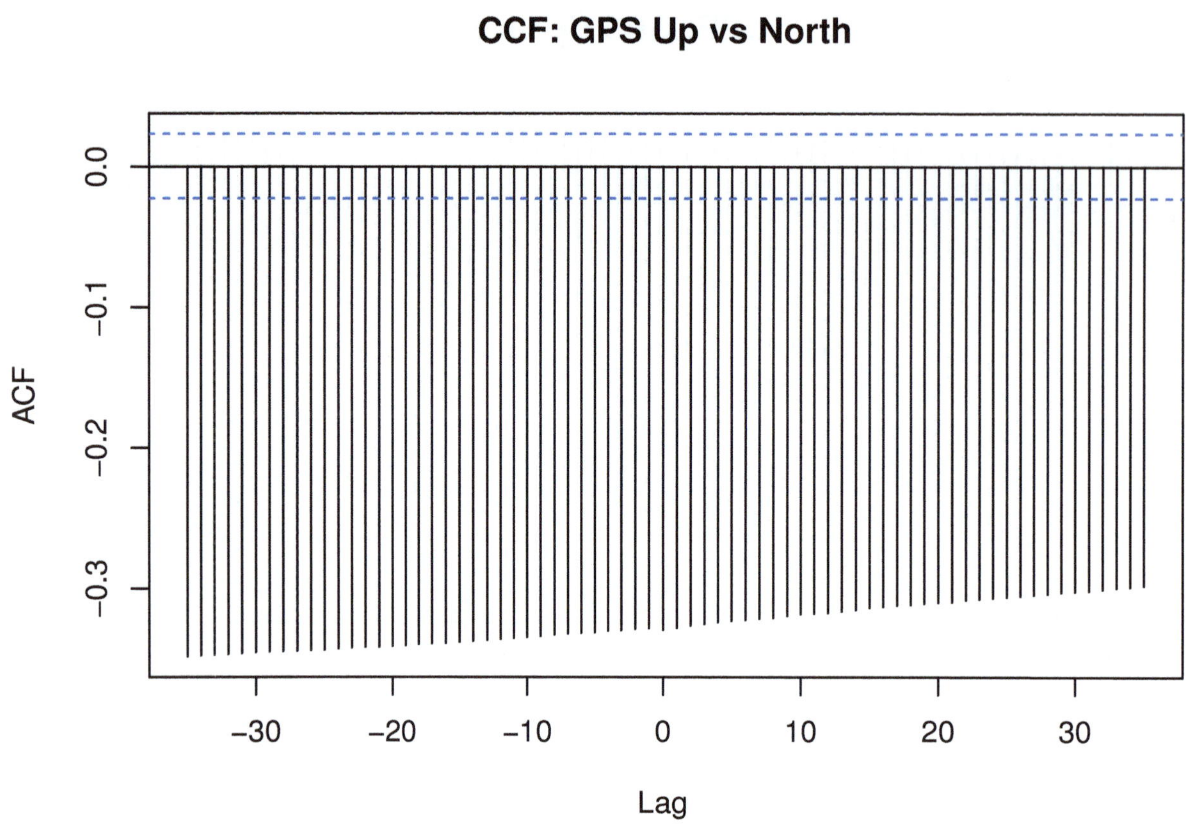

Figure 6.6: Cross-correlation between vertical (Up) and northward displacement.

Vertical and horizontal motion components are related (Figure 6.5, Figure 6.6). The **cross-correlation** (ccf) results show that vertical (up) and horizontal (east or north) displacements are related, suggesting that motion occurs in multiple directions simultaneously. However, the absence of a clear dominant lag indicates that these movements are not simply leading or lagging each other, but are likely driven by the same underlying processes. This supports the idea that tectonic deformation acts in several directions at once.

> **i** Pause and Connect the Ideas
>
> Until now, we examined the signal as a whole.
> Next, we begin separating the signal into parts: - long-term trend, - repeating seasonal behavior, - and short-term variability.
> This helps us understand which processes dominate the data.

6.4 Separating the Signal into Components

Decomposition splits one complicated signal into three simpler pieces: the slow trend, the repeating seasonal cycle, and the leftover (the remainder). Studying each piece on its own helps distinguish long-term deformation from repeating environmental effects and short-term variability.

> ⚠ Think before detrending
>
> Removing only the trend does not isolate the underlying process. Seasonal effects can still dominate the signal, and ignoring them may lead to incorrect interpretations. Both trend and seasonality must be considered together.

> Why Separate the Components?
>
> Imagine trying to understand a person's weight change over time. The long-term trend may show gradual weight gain or loss, while seasonal effects could reflect holiday eating patterns. If you only look at the trend, you might miss the fact that weight fluctuates seasonally, which is an important part of the overall picture. Another example is with a song made of three parts: a slow background melody, a repeating rhythm, and small variations in sound. To understand the song fully, you need to separate these elements rather than listening to everything at once.

We now apply a decomposition method that separates the signal into trend, seasonal behavior, and remaining variability.

```r
# stl() does the splitting. ts(..., frequency = 365) tells R the data are daily
# with a yearly cycle (approximately 365 days); s.window = "periodic" keeps
# that yearly shape steady.
decomp <- stl(ts(gps_cleaned$up, frequency = 365),
              s.window = "periodic")
plot(decomp)
```

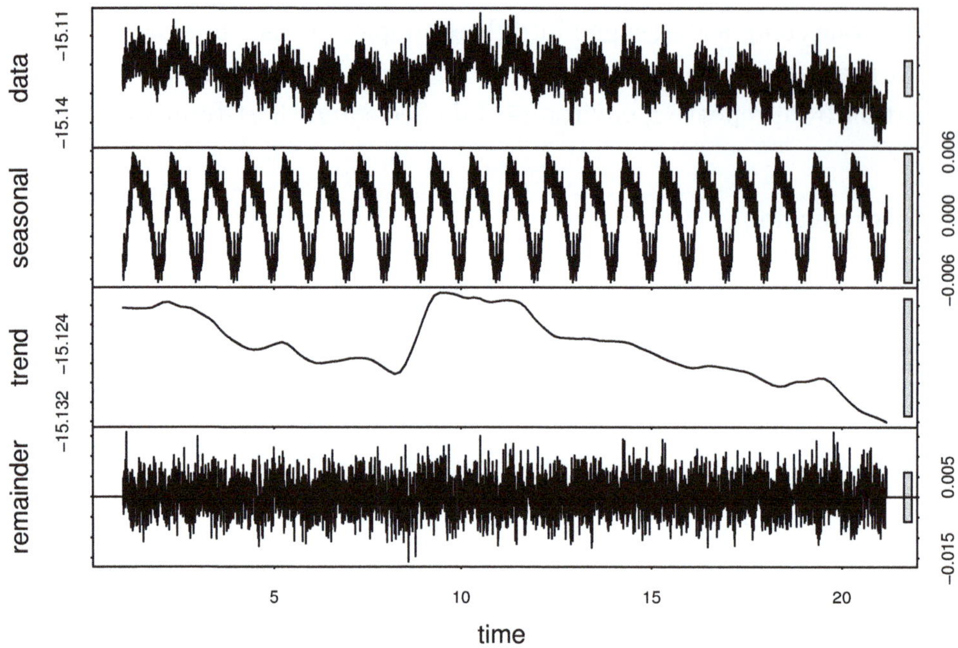

Figure 6.7: Time series decomposition of GPS displacement into trend, seasonal, and remainder components.

Interpretation

The decomposition plot (Figure 6.7) helps us examine the major parts of the signal separately. It stacks four panels: the original `data` on top, then `seasonal`, `trend`, and `remainder` beneath, and adding the three lower panels back together reproduces the data panel on top. The gray bar on the right of each panel marks the same vertical size, so a panel with a tall gray bar holds a small component and a short bar holds a large one.

> **i Understanding the decomposition plot**
>
> The decomposition separates the signal into three parts:
>
> - Trend → long-term behavior
> - Seasonal → repeating cycles
> - Remainder → variability not explained by the other two components
>
> Instead of studying one complex signal, we now study each part separately.

6.4.1 Understanding the Components

The **seasonal** component confirms (Figure 6.7) the repeating annual behavior identified earlier in the periodogram. The stability of this cycle suggests that seasonal influences, such as temperature changes or hydrological loading, affect the GPS signal in a predictable way each year.

The **trend** component captures long-term changes in displacement after removing seasonal effects. This smoother curve represents slow long-term changes in how the Earth's surface moves through time.

The **remainder** component contains short-term variability that is not explained by either the trend or the seasonal cycle. The remainder appears more irregular, suggesting that much of the repeating and long-term structure has already been removed from the signal.

After removing the long-term trend, the remaining signal highlights short-term variability and seasonal behavior more clearly. The persistence of structured fluctuations indicates that seasonality remains a dominant component of the GPS signal, even after detrending.

Together, these results emphasize the importance of separating trend and seasonal components when interpreting geophysical time series.

Important insight

Detrending alone is not sufficient to isolate tectonic signals, as seasonal effects still remain. Both trend and seasonal components must be considered separately for proper interpretation and isolation of geological or other underlying processes. This analysis shows that GPS displacement reflects a combination of long-term tectonic motion and seasonal environmental effects. Distinguishing between these components is essential for correctly interpreting geophysical signals.

6.5 Before vs After: Isolating the Signal

6.5.1 Removing the Trend (Detrending)

Before interpreting the data, it is important to understand how preprocessing changes what we see. The original time series contains a combination of long-term trends, seasonal cycles, and short-term variability. By removing selected components, we can isolate specific behaviors and better understand the underlying processes. Comparing the original and processed signals allows us to see what has been removed and what remains, helping us focus on the part of the signal most relevant to our question.

```
# Detrend: subtract the long-term trend, leaving the seasonal cycle and
# short-term variation.
# as.numeric() keeps the result a plain number series, so we can plot it
# against dates.
gps_detrended <- gps_cleaned$up - as.numeric(decomp$time.series[, "trend"])
```

```r
par(mfrow=c(1,2))
plot(gps_cleaned$date, gps_cleaned$up, type = "l",
     col = "blue", lwd=0.3,
     main = paste("Cleaned & Interpolated: \nGPS - Up"),
     xlab = "Date", ylab = "Up")

plot(gps_cleaned$date, gps_detrended, type = "l",
     col = "blue", lwd=0.3,
     main = paste("Detrended \nGPS Up"),
     xlab = "Date", ylab = "Up (Detrended)")
```

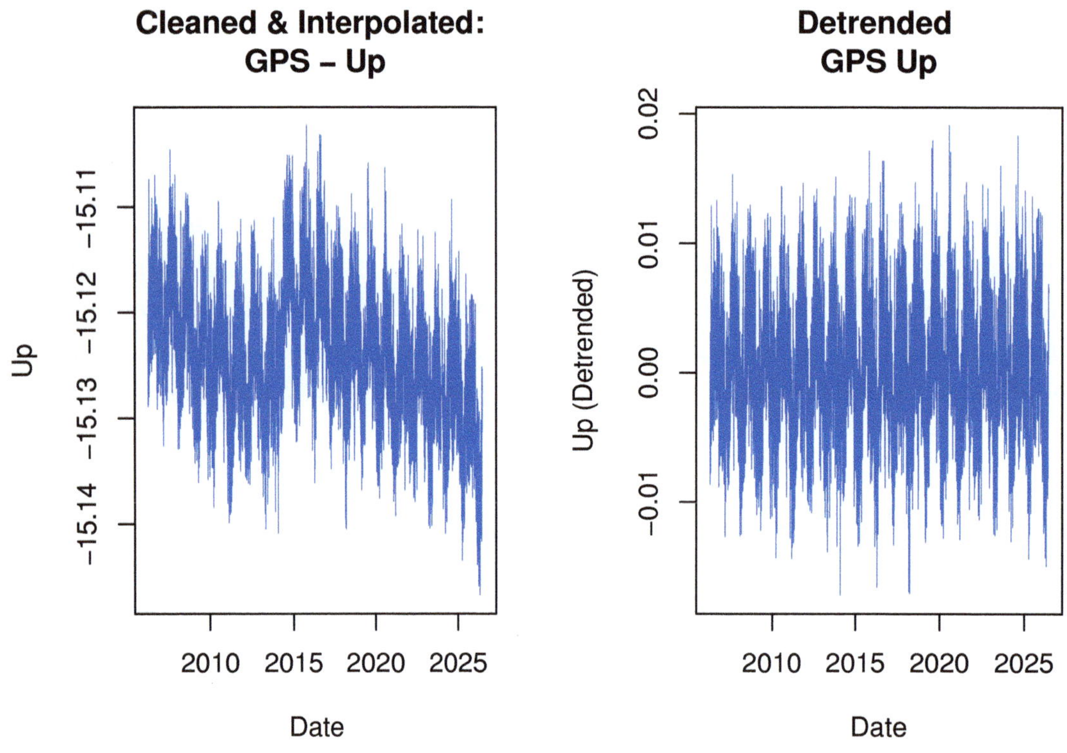

Figure 6.8: Comparison of the cleaned GPS signal and the detrended series, after removing the long-term trend.

The left panel shows the cleaned signal, while the right panel shows the same data after removing the long-term trend, which makes the seasonal cycle and short-term variability easier to see (Figure 6.8).

⚠ Think before removing

Removing components such as trend or seasonality simplifies the signal, but it also removes information. If done without clear reasoning, important physical processes may be lost. Always be explicit about what you remove and why, and interpret the results in that context.

Removing Components Changes the Signal

Imagine listening to a piece of music where drums, melody, and vocals are all playing together. If you remove the rhythm, the song may sound smoother, but you have also removed an essential part of its structure. Similarly, removing trend or seasonality can clarify the signal, but it changes what you are observing.

```
# Overlay: Original vs Seasonal vs Trend
# Instead of separate panels, let's see them on the same timeline.
season <- decomp$time.series[, "seasonal"]
trend  <- decomp$time.series[, "trend"]

plot(gps_cleaned$date, gps_cleaned$up - mean(gps_cleaned$up, na.rm = TRUE),
     type = "l", col = "gray",
     xlab = "Date", ylab = "Centered displacement",
     main = "GPS Up with Trend and Seasonal Components")

lines(gps_cleaned$date, trend - mean(gps_cleaned$up, na.rm = TRUE),
      col = "red", lwd = 2)

lines(gps_cleaned$date, season,
      col = "blue", lwd = 1)
legend("topright",
       legend = c("Original (centered)", "Trend", "Seasonal"),
       col = c("gray", "red", "blue"),
       lty = 1,
       lwd = c(1, 2, 1),
       bty = "o",          # box around legend (not transparent)
       bg = "white",       # solid white background
       cex = 0.8,          # smaller text size
       inset = 0.02)       # slight margin from edge
```

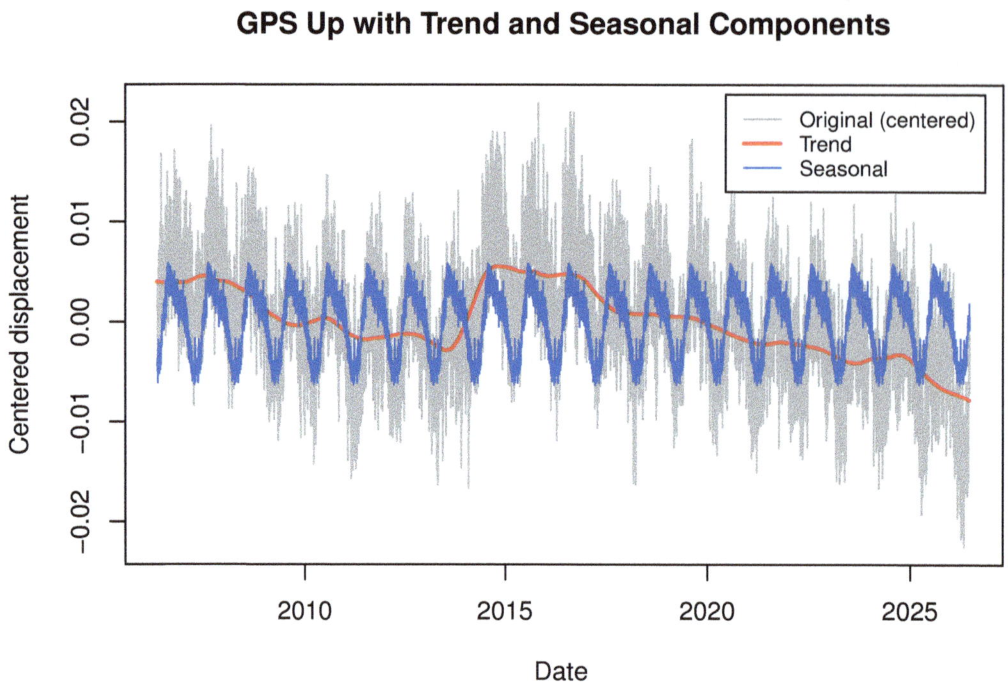

Figure 6.9: Overlay of original, trend, and seasonal components of GPS displacement.

Trend and seasonal components combine to produce the observed signal (Figure 6.9). The seasonal component is small in amplitude relative to the full GPS signal, but its repeated annual pattern is clear. This shows that even subtle oscillations can be scientifically important when they recur consistently over time.

6.5.2 Removing the Seasonality (Deseasoning)

Now we instead remove the seasonal cycle, leaving the long-term trend and short-term variation.

```
# Deseason: subtract the seasonal cycle.
# as.numeric() keeps the result a plain number series, matching the plots
# that follow.
gps_deseason <- gps_cleaned$up - as.numeric(decomp$time.series[, "seasonal"])
```

```
par(mfrow=c(1,2))
plot(gps_cleaned$date, gps_cleaned$up, type="l",
     lwd = 0.2,
     main="Original \nGPS Signal",
     ylab="Up")
```

```
plot(gps_cleaned$date, gps_deseason, type="l",
     lwd = 0.2, col = "darkgreen",
     main = "After Removing \nSeasonal Component",
     ylab = "Up (Deseasonalized)")
```

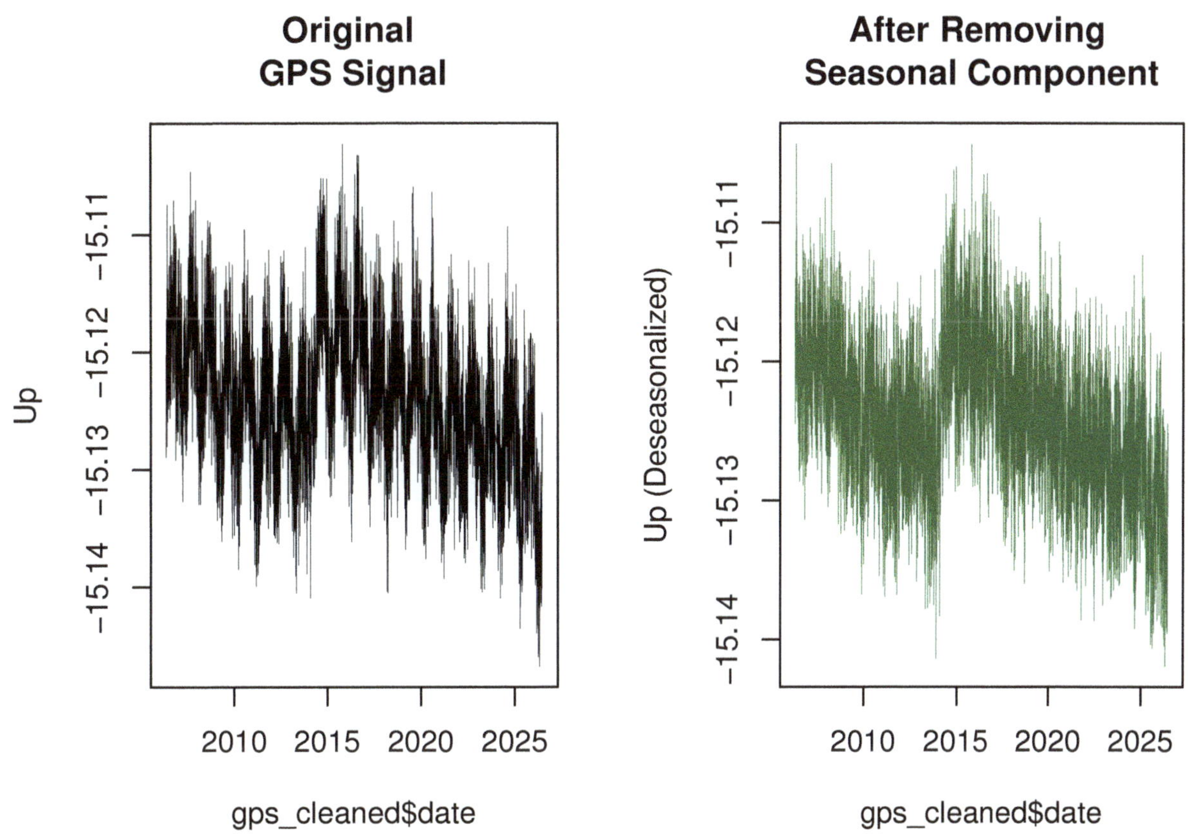

Figure 6.10: GPS vertical (Up) signal before and after removing the seasonal component.

```
plot(gps_cleaned$date, gps_cleaned$up, type="l",
     lwd = 0.2,
     main="Original vs Deseasonal \nGPS Signal",
     ylab="Up")
lines(gps_cleaned$date, gps_deseason, type = "l",
      lwd = 0.2, col = "darkgreen",
     main = "After Removing \nSeasonal Component",
     ylab = "Up (Deseasonalized)")
```

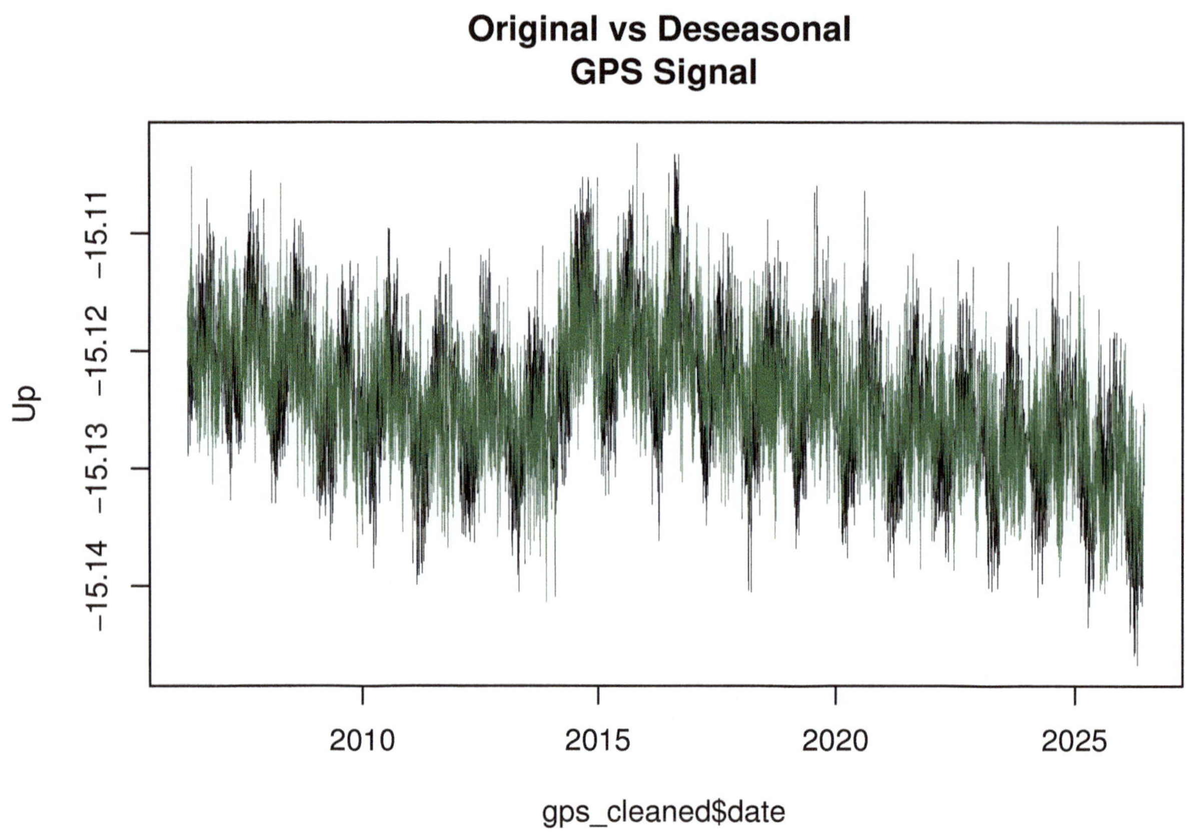

Figure 6.11: Original and deseasonalized GPS vertical (Up) signal shown together for comparison.

Interpretation

After removing the seasonal component, the oscillating pattern disappears, revealing the underlying long-term behavior more clearly. Surprisingly, the signal appears more irregular and noisier. This is because the smooth, repeating pattern has been removed, leaving behind short-term variability and noise. Although the series looks more "noisy," this does not necessarily mean that variability has increased. Instead, the structured seasonal variation has been removed, making the remaining fluctuations more visible.

```
cat(paste0("Checking variance before and after removing seasonality:\n"))
```

```
Checking variance before and after removing seasonality:
```

```
var_original <- var(gps_cleaned$up, na.rm=TRUE);var_original
```

```
[1] 3.948678e-05
```

```r
var_deseason <- var(gps_deseason, na.rm=TRUE);var_deseason
```

```
[1] 2.825035e-05
```

Variance: The reduction in variance after removing the seasonal component indicates that a significant portion of the variability in the original signal was driven by periodic processes.

Confirming the Removal of Seasonality

The dominant annual peak is gone, which shows the strong periodic behavior was seasonal.

By removing this seasonal component, the underlying trend becomes more clearly visible, allowing us to focus on long-term deformation that is more likely associated with tectonic processes.

What Changed After Removing Seasonality

After removing the seasonal cycle:

- the annual repeating pattern becomes weaker,
- short-term variability becomes easier to see,
- and the remaining signal better highlights long-term deformation processes.

To further examine what remains after removing seasonality, we now shift from viewing the signal through time to examining the strength of repeating cycles.

Periodogram after Removing Seasonality

```r
par(mfrow = c(1,2))
plot(1/sp$freq, sp$spec, type = "l",
     main = "Periodogram of\n Original GPS Up",
     cex.main = 0.8,
     xlab = "Period (in days)",
     ylab = "Spectral Density",
     xlim=c(0,500))
sp2 <- TSA::periodogram(gps_deseason, plot = FALSE)
plot(1/sp2$freq, sp2$spec, type="l", col = "darkgreen",
     main = "Periodogram After\n Removing Seasonality", cex.main=0.8,
     xlab = "Period (in days)",
     ylab = "Spectral Density",
     xlim = c(0,500),
     # Keep same y-axis scale for comparison
     ylim = c(0, max(sp$spec, na.rm=TRUE)))
```

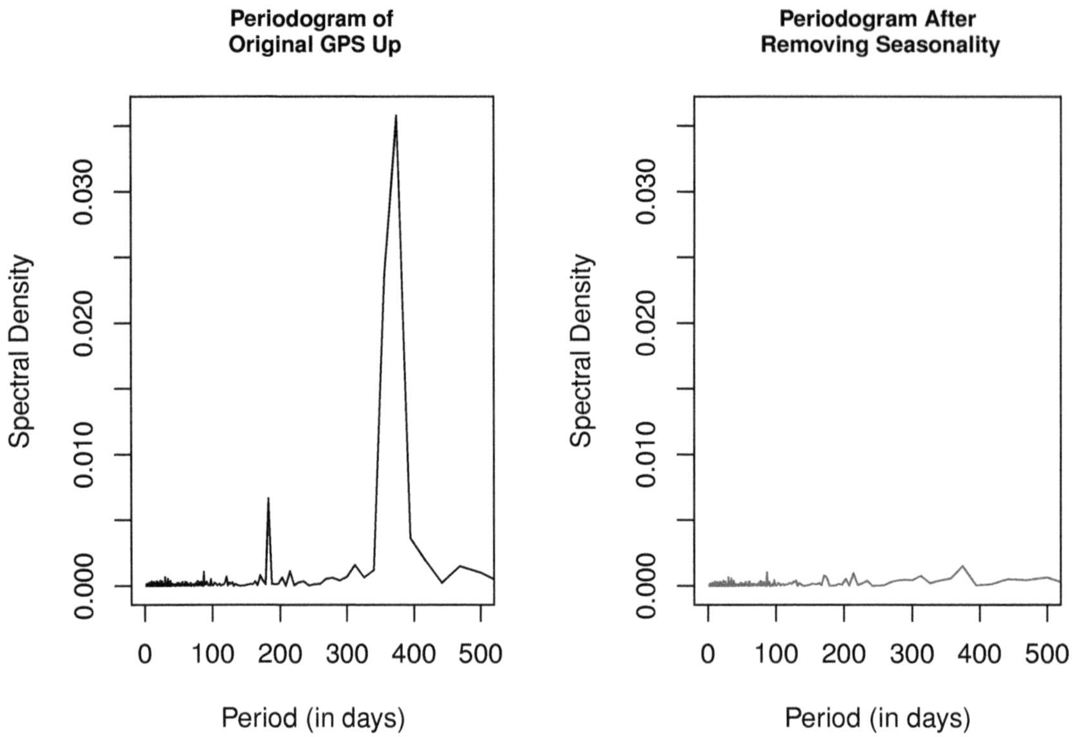

Figure 6.12: Comparison of GPS Up (vertical) periodograms before and after removing seasonal components.

The periodogram (Figure 6.12) after removing seasonality shows that the strong peak at the annual period has disappeared, confirming that the dominant periodic signal was due to seasonal effects. This reinforces the importance of separating components to avoid misinterpreting environmental noise as tectonic signals.

Autocorrelation after Removing Seasonality

Before examining the plots, it is important to consider how removing seasonality may affect the autocorrelation structure. The original ACF was strongly influenced by a periodic annual signal, which introduces repeating patterns and creates high correlations at specific lags. Once this seasonal component is removed, the ACF may change in several ways:

- **Decrease in autocorrelation**
  This is the most common outcome. If the original signal was dominated by a periodic cycle, removing it will reduce correlations across many lags.
  → Interpretation: much of the structure was driven by external, repeating signals (e.g., environmental or climatic effects).
- **Persistence remains at short lags**
  Autocorrelation may still be strong at small lags even after removing seasonality.

→ Interpretation: the system has intrinsic memory, meaning current values depend on recent past values (e.g., tectonic deformation or gradual physical processes).

- **Increase in autocorrelation (less common)**
  In some cases, correlations at short lags may become more pronounced after removing seasonality.
  → Interpretation: the seasonal signal was masking an underlying process. Once removed, the internal dynamics (e.g., slow drift or deformation) become clearer.
- **Change in decay pattern**
  The rate at which autocorrelation decreases with lag may change.
  → Interpretation: This reflects a shift from periodic behavior to more irregular or trend-driven behavior.

> **!** Separating Different Processes
>
> Removing seasonality helps separate: - externally driven periodic signals, and
> - internally driven system dynamics.

```
par(mfrow = c(1,2))
acf(gps_cleaned$up, main = "ACF of \nCleaned GPS Up")
acf(gps_deseason, main = "ACF after \nRemoving Seasonality")
```

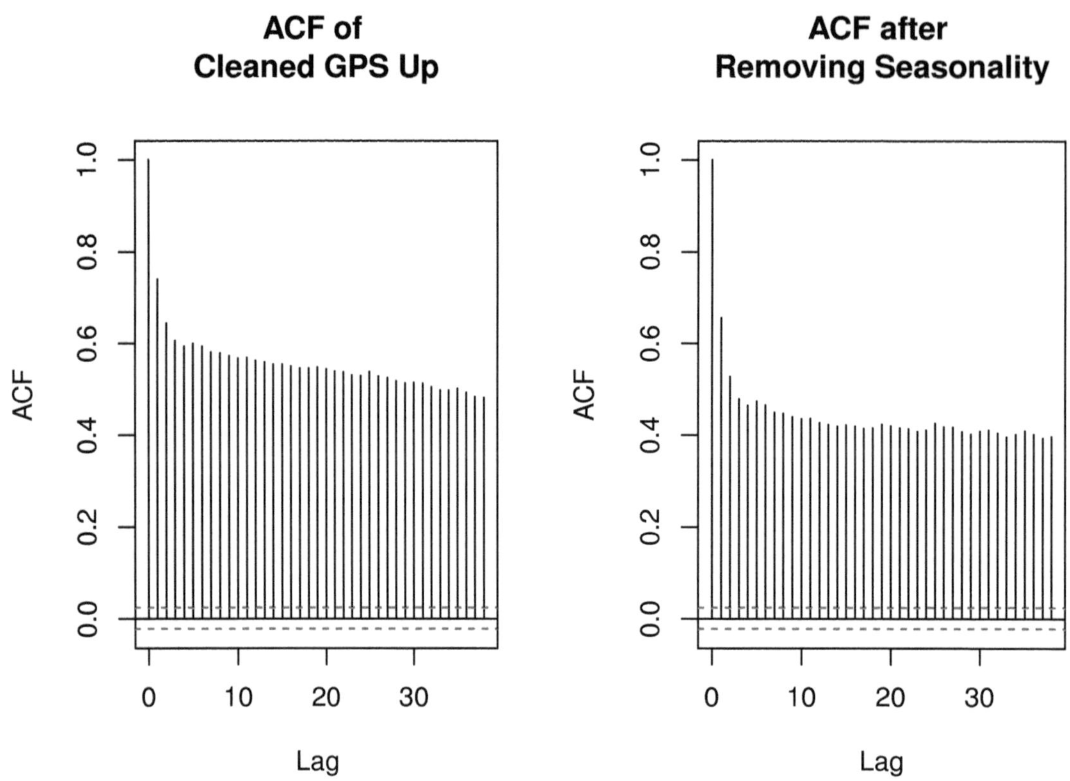

Figure 6.13: Autocorrelation comparison before and after removing seasonal components.

After removing seasonality, the ACF shows (Figure 6.13) a reduction in overall correlation compared to the original series, indicating that a substantial portion of the structure was driven by the seasonal component. However, strong autocorrelation persists at short lags, demonstrating that the system retains memory and that displacement is not random. This remaining persistence likely reflects underlying physical processes, such as gradual tectonic deformation, rather than periodic environmental forcing. This is what we would expect if the annual cycle was a dominant repeating signal in the data. After removing it, what remains are the slower and more persistent processes that still control how the system evolves over time.

6.5.3 Thinking Like a Scientist: A Time Series Workflow

This case study presents a real analysis workflow with an emphasis on interpretation. Simple methods are used to answer specific questions about the data, allowing us to focus on understanding rather than complexity. The goal is to demonstrate how a scientist approaches real data: by clearly defining the question, selecting an appropriate method, and interpreting the results in a way that connects back to the original problem.

All methods are intentionally chosen to be accessible and directly tied to the questions we want to answer, rather than relying on more complex techniques that may obscure interpretation. The key is not to use more advanced methods, but to use the right method for the question and to interpret the results meaningfully.

Each method is used to answer a specific question:

Method	Question answered
Plot	What is happening overall?
Moving Average	What is the underlying long-term trend?
ACF	Does the system have memory or persistence?
CCF	Are two variables related, and is there a time lag between them?
Periodogram	Are there repeating cycles (frequencies)?

Together, these methods provide complementary perspectives (time domain, frequency domain, and relationships between variables), allowing us to build a complete understanding of the system. You can reuse the same pattern for other case studies:

- **Temperature**: plot + moving average + interpretation of seasonality and trend
- **Pollution**: map + spatial interpretation
- **River flow**: lag-based model + interpretation of system memory

The key is to understand that a scientist thinks first through real data using simple, meaningful tools. More advanced methods can be added later, but the foundation is to ask clear questions and use appropriate methods to answer them.

The same workflow can be adapted to other geoscience problems by choosing the right data and the right question. The first case study focused on time series analysis. We now shift to a spatial application, where elevation data is used to explore flood exposure.

! What We Learned from Case Study 1

In the GPS example, we learned how to:

- clean and prepare a time series,
- identify repeating patterns,
- examine system memory,
- separate trend and seasonality,
- and interpret how preprocessing changes the signal.

The next case study shifts from time series analysis to spatial analysis, where terrain and elevation help explain flood exposure.

6.6 Case Study 2: Flood Exposure and Coastal Vulnerability in Long Island, NY

This case study returns to the Long Island DEM from Chapter 3 (a Digital Elevation Model, which is simply a map of ground height at every point) and uses it to explore how terrain controls flood exposure under different water levels.

> **ℹ Note for beginners**
>
> The code that follows is more advanced, including custom functions and additional visualization packages (ggnewscale, tidyterra, patchwork). You do not need to understand every line of code. Focus on reading the maps and the interpretation sections. If you want to run the code, install the new packages first by running:
> `install.packages(c("ggnewscale", "tidyterra", "patchwork"))`

```r
# Core spatial packages
library(sf)
library(terra)
library(tigris)
library(elevatr)
library(ggplot2)
library(tmap)

# Optional packages used for cleaner map styling and area summaries
library(units)
library(patchwork)
library(grid)
library(ggnewscale)
library(tidyterra)

# Keep tigris and tmap stable for reproducible output
options(tigris_use_cache = TRUE)
tmap_mode("plot")


# Goal of this block: download the four Long Island counties and merge them
# into
# one clean outline of the island that we can place other maps on top of later.
# Download the four Long Island counties directly from the NYS civil
# boundaries service
query <- paste0(
```

```
"https://gisservices.its.ny.gov/arcgis/rest/services",
  "/NYS_Civil_Boundaries/MapServer/3/query?",
  "where=",
  URLencode("NAME IN ('Kings','Queens','Nassau','Suffolk')",
            reserved = TRUE),
  "&outFields=NAME,ABBREV",
  "&returnGeometry=true",
  "&f=geojson"
)
# Read the Long Island county polygons
li_parts <- st_read(query, quiet = TRUE)

# Fix invalid geometries
li_parts <- st_make_valid(li_parts)

# Switch to UTM Zone 18N (code 32618): a map system that measures in meters,
# which makes area and distance calculations for this region accurate.
li_parts <- st_transform(li_parts, 32618)

# Dissolve all counties into one polygon
long_island <- st_union(li_parts)

# Keep only polygon geometry
long_island <- st_collection_extract(long_island, "POLYGON")

# Fix geometry again after dissolving
long_island <- st_make_valid(long_island)
```

The Long Island county boundaries are downloaded from the New York State GIS Program Office (the NYS Civil Boundaries service).

Before simulating flooding, we first need elevation data that describe the shape and height of the terrain. These data are downloaded with the `elevatr` package, which retrieves public AWS Terrain Tiles.

6.6.1 Step 1: Download and reproject the DEM for analysis

```
# Download the elevation grid. A raster is a grid of pixels, each holding
# one value
# (here, ground height). z = 10 sets the detail level (higher z = finer
# pixels).
```

```
dem <- get_elev_raster(
  locations = st_transform(st_as_sf(long_island), 4326),
  z = 10,
  clip = "locations"
)

dem <- rast(dem)
dem_utm <- project(dem, "EPSG:32618")
li_v <- vect(st_transform(st_as_sf(long_island), 32618))

# Extract raster values inside Long Island

elev <- terra::extract(dem_utm, li_v)
elev <- elev[!is.na(elev[, 2]), ]
summary(elev[, 2])
```

```
   Min. 1st Qu.  Median    Mean 3rd Qu.    Max.
 -88.75    6.00   16.41   21.35   32.00  173.45
```

Interpretation: This summary tells us the range of elevation values inside Long Island. Low values are the most important for flooding.

6.6.2 Step 2: Compute flooded percentage for several water levels

```
levels <- c(0, 5, 10, 15, 20, 25, 30)

results <- data.frame(
  level = levels,
  flooded_cells = NA,
  percent = NA
)

total_cells <- nrow(elev)

for (i in seq_along(levels)) {
  lvl <- levels[i]
  flooded_cells <- sum(elev[, 2] <= lvl, na.rm = TRUE)
  results$flooded_cells[i] <- flooded_cells
  results$percent[i] <- 100 * flooded_cells / total_cells
}
```

```
results
```

```
  level flooded_cells   percent
1     0         33265  2.985773
2     5        252782 22.688999
3    10        394998 35.453907
4    15        524431 47.071448
5    20        636493 57.129817
6    25        726725 65.228787
7    30        807808 72.506568
```

Interpretation: This table gives a simple estimate of how much of Long Island lies below each water level. It is a simplified first estimate, not a complete flood simulation.

6.6.3 Step 3: Plot the flooding trend

```
ggplot(results, aes(x = level, y = percent)) +
  geom_line(linewidth = 1) +
  geom_point(size = 2) +
  theme_minimal() +
  labs(
    title = "Projected Flooding of Long Island",
    x = "Sea-Level Rise (m)",
    y = "% Land Submerged"
  )
```

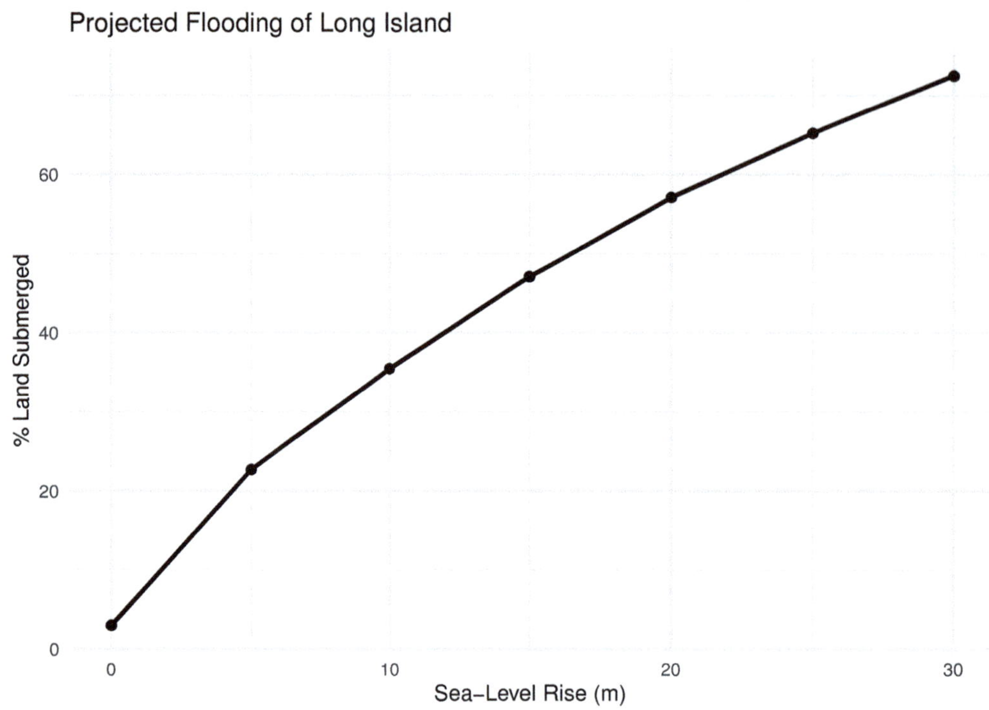

Figure 6.14: Percentage of land submerged as a function of sea-level rise.

Interpretation: The line (Figure 6.14) should rise as water level increases. In a low-lying coastal setting, small increases in water level can produce large changes in flooded area.

6.6.4 Step 4: Show a flood map at a chosen water level

```
# Mark every pixel at or below 10 m as flooded (value 1); leave the rest
# empty (NA).
# This is our simple "if the water reaches 10 m, what goes under?" map.
flood10 <- ifel(dem_utm <= 10, 1, NA)
flood10 <- mask(flood10, li_v)

# Check whether anything is actually flooded
global(flood10, "sum", na.rm = TRUE)
```

```
                      sum
file679c7d7b3d89 416831
```

```
global(!is.na(flood10), "sum", na.rm = TRUE)
```

```
                      sum
```

```
file679c7d7b3d89 416831
```

```
# A hillshade is a shaded-relief picture that makes flat terrain look 3D,
# as if lit
# by the sun, so the map reads naturally. The next lines build that shaded
# background.
# We exaggerate elevation first to make the gentle coastal relief easier to
# see.
dem_ex <- dem_utm * 5
# Compute terrain properties
slope  <- terrain(dem_ex, "slope", unit = "radians")
aspect <- terrain(dem_ex, "aspect", unit = "radians")
# Create nice lighting for perception
hillshade <- shade(
  slope, aspect,
  angle = 40,
  direction = 315
)

# Contrast stretch
hs_vals <- values(hillshade)
q <- quantile(hs_vals, probs = c(0.02, 0.98), na.rm = TRUE)

hillshade_stretch <- clamp(hillshade, q[1], q[2])
hillshade_stretch <- (hillshade_stretch - q[1]) / (q[2] - q[1])

hillshade_utm <- project(hillshade_stretch,
                         dem_utm,
                         method = "bilinear")

hillshade_utm <- crop(hillshade_utm, ext(li_v))

plot(
  hillshade_utm,
  col = gray(seq(0.15, 0.95, length.out = 100)),
  legend = FALSE,
  axes = FALSE,
  box = FALSE
)
plot(
  flood10,
  add = TRUE,
```

```
  col = adjustcolor("deepskyblue", alpha.f = 0.5),
  legend = FALSE
)
plot(li_v, add = TRUE, border = "black", lwd = 0.6)
```

Figure 6.15: Simulated flood extent over Long Island at a selected water level, overlaid on hillshade.

Interpretation: Blue areas (Figure 6.15) mark land that would fall below the selected water level. The hillshade helps the viewer understand the terrain context behind the flood pattern.

6.6.5 Step 5: Compare several flood levels in a professional figure

```
# Helper for a clean scale bar
nice_scale <- function(target_m) {
  p10 <- 10^floor(log10(target_m))
  choices <- c(1, 2, 5) * p10
  choices[which.min(abs(choices - target_m))]
}
# This helper builds one finished flood map for a given water level. You do
# not need
```

```
# to follow every line, just know that calling make_panel(5) returns the 5
# m map.
make_panel <- function(level_m, add_scale_north = FALSE) {
  bb <- ext(hillshade_utm)
  w  <- bb$xmax - bb$xmin
  h  <- bb$ymax - bb$ymin

  flood <- ifel(dem_utm <= level_m, 1, NA)
  flood <- mask(flood, li_v)

  p <- ggplot() +
    geom_spatraster(data = hillshade_utm) +
    scale_fill_gradientn(colors = gray(seq(0.15, 0.95, length.out = 100)),
                         guide = "none") +
    ggnewscale::new_scale_fill() +
    geom_spatraster(data = as.factor(flood), alpha = 0.55) +
    scale_fill_manual(values = c("1" = "deepskyblue"),
                      na.value = NA, guide = "none") +
    coord_sf(crs = st_crs(32618), expand = FALSE) +
    scale_x_continuous(n.breaks = 2) +
    scale_y_continuous(n.breaks = 4) +
    theme_minimal(base_size = 11) +
    theme(
      axis.ticks = element_line(linewidth = 0.3, color = "gray50"),
      axis.ticks.length = unit(0.15, "cm"),
      axis.text = element_text(size = 8, color = "gray50"),
      panel.grid = element_blank(),
      axis.title = element_blank(),
      plot.title = element_text(face = "bold", hjust = 0.5)
    ) +
    labs(title = paste0(level_m, " m"))

  if (add_scale_north) {
    scale_target <- 0.22 * w
    scale_m <- nice_scale(scale_target)

    nx  <- bb$xmin + 0.08 * w
    ny1 <- bb$ymax - 0.08 * h
    ny0 <- ny1 - 0.10 * h

    x1 <- bb$xmax - 0.05 * w
```

```
    x0 <- x1 - scale_m
    y0 <- bb$ymin + 0.09 * h

    scale_label <- if (scale_m >= 1000) {
      paste0(scale_m / 1000, " km")
    } else {
      paste0(scale_m, " m")
    }

    p <- p +
      annotate(
        "segment",
        x = x0, xend = x1,
        y = y0, yend = y0 + 0.005 * h,
        colour = "gray40", linewidth = 0.7
      ) +
      annotate(
        "text",
        x = (x0 + x1) / 2,
        y = y0 - 0.04 * h,
        label = scale_label,
        colour = "gray40", size = 3
      ) +
      annotate(
        "segment",
        x = nx, xend = nx,
        y = ny0 - 0.01 * h, yend = ny1 - 0.02 * h,
        arrow = arrow(length = unit(0.18, "cm")),
        colour = "gray40", linewidth = 0.4
      ) +
      annotate(
        "text",
        x = nx,
        y = ny1 + 0.02 * h,
        label = "N",
        colour = "gray40", size = 3,
        fontface = "plain"
      )
  }
  p
}
```

```
p0  <- make_panel(0,  add_scale_north = TRUE)
p5  <- make_panel(5)
p10 <- make_panel(10)
p15 <- make_panel(15)

fig <- (p0 | p5) / (p10 | p15)
fig
```

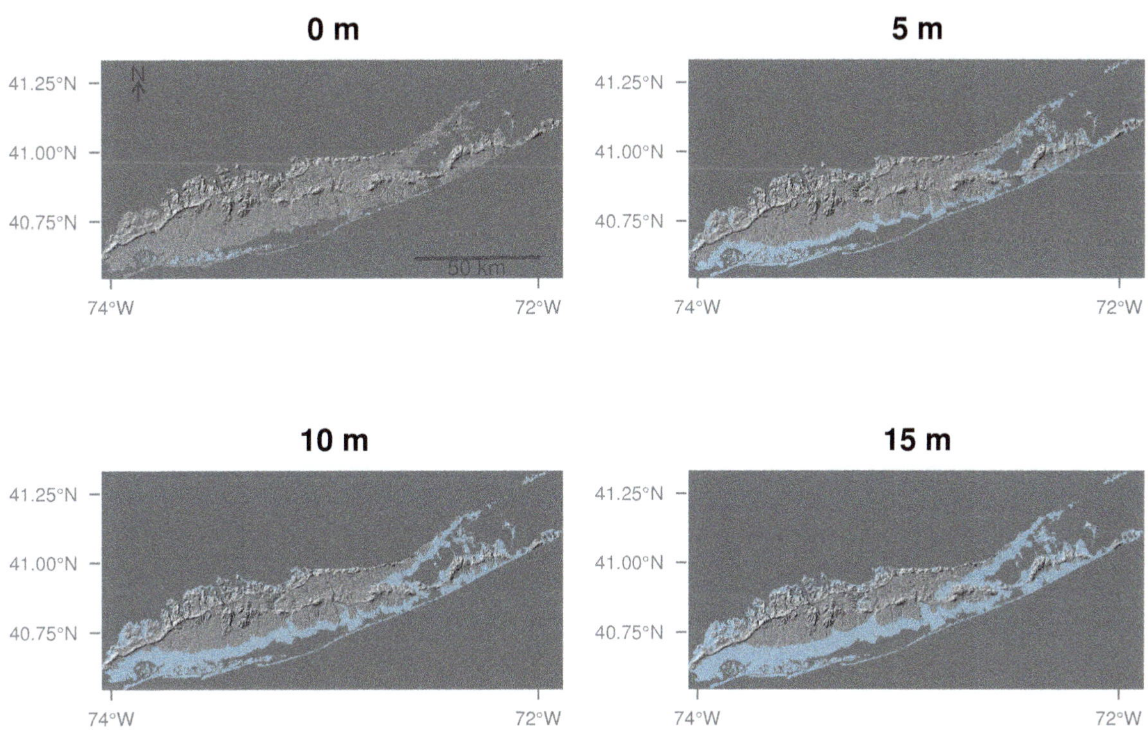

Figure 6.16: Comparison of flood extent on Long Island, NY under multiple sea-level scenarios.

Interpretation: These panels (Figure 6.16) make the flood progression easy to compare. As the water level increases, the flooded area expands.

> Compare the Maps Side by Side
>
> Comparing several maps side by side makes changes in flooded areas much easier to detect.

Interpretation

The flood simulation shows how sensitive a low-lying coastal island can be to small changes in water level.

What the map tells us

- Areas that are most vulnerable to rising water levels
- As water level rises from 0 m to 5 m, the flooded area expands rapidly
- A barrier island can have large changes in inundation even for a small increase in water level

What the map does not tell us

- the timing of future flooding
- waves, storm surge dynamics, or tides
- erosion, dune failure, or water flow pathways
- drainage, levees, or protective infrastructure

It is a static elevation-threshold model, not a full hydrodynamic flood model.

This means the model is useful for learning and for identifying potentially vulnerable areas. However, it is not detailed enough for emergency planning or engineering decisions without more advanced analysis.

> Why Simple Models Matter
>
> A basic elevation-threshold model is not the final answer, but it is a very good evaluation tool. It helps us study how terrain controls vulnerability.

6.7 Case Study 3: Climate Change Effects on Ice Mass

6.7.1 Introduction

In the previous chapters, we learned how to work with spatial data, time series, and simple models. In this case study, we combine those ideas (time and space) to study a real climate problem:

How is ice mass changing over time, and what does that tell us about climate change?

> How GRACE Measures Ice Loss
>
> Imagine weighing a giant sheet of ice every month from space. If the weight keeps going down, something important is changing. Ice sheets are sensitive to temperature changes, and if ice melts, then mass decreases. GRACE allows us to measure those changes across very large regions.

Before looking at the code or maps, focus on the main scientific idea:

- if Greenland loses ice, mass decreases,
- if ice increases, mass increases,
- and GRACE helps us measure those large-scale changes from space.

Here we use data from the GRACE (Gravity Recovery and Climate Experiment) satellite mission. GRACE measures changes in Earth's mass from space using two satellites that follow each other along the same orbit. Mass changes on Earth, such as melting ice or shifting water above or below the ground, cause tiny variations in the gravitational field. More mass produces slightly stronger gravity, while less mass produces slightly weaker gravity. As those two satellites travel around the Earth, they continuously measure the distance between them with extremely high precision (microns). When the leading satellite passes over a region with slightly stronger gravity, it is pulled forward, increasing the distance between the two satellites. About 29 seconds later, the second satellite reaches the same region, experiences the same pull, and the distance changes again.

In short: more mass below means stronger gravity, which nudges the two satellites slightly farther apart, and GRACE turns that tiny spacing change into a measurement of where mass on Earth is increasing or decreasing through time, including changes in ice, water, and other large-scale movements.

When ice melts, mass is redistributed as water moves away from the ice sheet, slightly weakening the local gravitational field. GRACE detects these changes through time, allowing us to monitor how ice sheets such as Greenland evolve. Because the solid Earth (such as mountains and crust) does not change significantly over short timescales, most of the observed variation comes from moving mass, including water, ice, and atmospheric changes. In Greenland, where cold temperatures limit groundwater movement, the dominant signal is related to changes in ice thickness.

GRACE measures large regional changes rather than small local details. This makes it useful for studying large ice sheets such as Greenland. The spatial resolution is about 300 km (approximately

3 degrees, where 1 degree is about 111 km). This means that each data point represents an average over a very large area rather than a precise local measurement. GRACE is therefore best suited for detecting broad regional changes rather than small-scale variations. Despite this limitation, it lets us monitor large-scale mass changes that would be hard to observe directly.

We will investigate how Earth's ice mass is changing over time and what that tells us about climate change. The key question is: Is Greenland losing ice over time, and how can we observe that change in both space and time?

6.7.2 Load the data and understand the variables

We begin by loading the packages needed to read the GRACE NetCDF file and work with dates. To keep the results in this chapter reproducible, we use a fixed teaching copy of the GRACE data stored on Zenodo, a public research data archive, under the permanent identifier DOI: 10.5281/zenodo.20535360. Zenodo gives the file a stable web address that does not change, so the code below can download exactly the same dataset at any time. The original data are produced by NASA's Jet Propulsion Laboratory and are freely available under NASA's open data policy. If you would like data more recent than the snapshot used here, you can download the latest official release from NASA's Physical Oceanography Distributed Active Archive Center (PO.DAAC). This requires a free NASA Earthdata Login account, which you can create at https://urs.earthdata.nasa.gov. The version used in this chapter is the JPL mascon CRI-filtered solution (Release 06.3, Version 04), available at https://podaac.jpl.nasa.gov/dataset/TELLUS_GRAC-GRFO_MASCON_CRI_GRID_RL06.3_V4. Full attribution is given in the References section.

```
# The GRACE data is stored on Zenodo (DOI: 10.5281/zenodo.20535360).
# Run these lines once to download it into your working folder as
# GRACE_data.nc.
# After it is saved, you can skip this step on future runs.
url <- paste0(
  "https://zenodo.org/records/20535360/files/",
  "GRCTellus.JPL.200204_202601.GLO.RL06.3M.MSCNv04CRI.nc?download=1"
)
download.file(url, "GRACE_data.nc", mode = "wb")


# A NetCDF (.nc) file is a common scientific format that stores a grid of
# values
# plus its longitude, latitude, and time labels. The lines below open it
# and extract
# each piece (lon, lat, time, and the mass values stored as "lwe_thickness").
library(ncdf4)
library(lubridate)
library(maps)
```

```
library(fields)

# Open the GRACE file you downloaded above
nc <- nc_open("GRACE_data.nc")

# Read the main variables
lon  <- ncvar_get(nc, "lon")
lat  <- ncvar_get(nc, "lat")
time <- ncvar_get(nc, "time")
mass <- ncvar_get(nc, "lwe_thickness")

# Convert the time variable to dates.
# Each dataset stores time as a count from its own chosen start date (the
# "origin").
# This GRACE file counts from 2002-01-01, while the GPS data earlier
# counted from
# 1858-11-17, so we pass the origin that matches this file. The different
# date is
# correct, not a typo.
date <- as.Date(time, origin = "2002-01-01")
```

The NetCDF file stores the data on a spatial grid. Each grid cell contains a value, and each time step represents a new map through time.

The `mass` variable is a 3D array with longitude, latitude, and time dimensions. Each value is the "equivalent water thickness" (in centimeters), that is, the gained or lost mass expressed as the depth of water it would equal, which is a standard way to report GRACE results.

This means that negative values indicate mass below the reference average, which is often associated with mass loss over time (e.g., ice melting), while a positive value indicates a gain in mass.

The `lon` and `lat` variables define the spatial grid, while the `time` variable provides the temporal dimension. By analyzing how the `mass` variable changes over time and across space, we can investigate trends in ice mass loss or gain, particularly in regions like Greenland that are sensitive to climate change. Spatial analysis allows us to see where changes are happening, while time series analysis helps us understand how those changes evolve over time.

6.7.3 Define the study area

For this case study, we focus on **Greenland**. It is one of the best places to study climate change because its ice sheet is large and sensitive to warming. We will define a simple box around Greenland to extract the relevant data. The box is defined by its minimum and maximum longitude and latitude.

```
# This file numbers longitude from 0 to 360 (instead of -180 to 180), so
# Greenland's
# usual -75 to -10 becomes 285 to 350 here. We convert back to the familiar
# range later.
lon_min <- 285
lon_max <- 350
lat_min <- 58
lat_max <- 85

# Find the grid cells inside the Greenland box
lon_idx <- which(lon >= lon_min & lon <= lon_max)
lat_idx <- which(lat >= lat_min & lat <= lat_max)
```

This box is a simple region of interest. It is not a perfect political or glaciological boundary, but it is good enough for an introductory workflow. The grid cells that fall within this box will be used to compute the average ice mass change for Greenland over time. This time span allows us to analyze trends and seasonal patterns in ice mass change, which can provide insights into the effects of climate change on the Greenland ice sheet. The number of time steps in the `mass` variable equals the length of the `date` vector. Each time step represents a monthly measurement, so we can analyze how ice mass changes on a monthly basis over the entire period covered by the GRACE data.

6.7.4 Show where Greenland is located

A map helps us see the study area before looking at the data values. This is an important first step in spatial analysis.

```
world <- map_data("world")

ggplot() +
  geom_polygon(data = world,
               aes(x = long, y = lat, group = group),
               fill = "gray90", color = "gray40", linewidth = 0.2) +
  geom_rect(aes(xmin = -75, xmax = -10, ymin = 58, ymax = 85),
            fill = NA, color = "black", linewidth = 1.2) +
  coord_fixed(xlim = c(-100, 20), ylim = c(40, 90)) +
  labs(title = "Greenland Study Region",
       x = "Longitude", y = "Latitude") +
  theme_minimal()
```

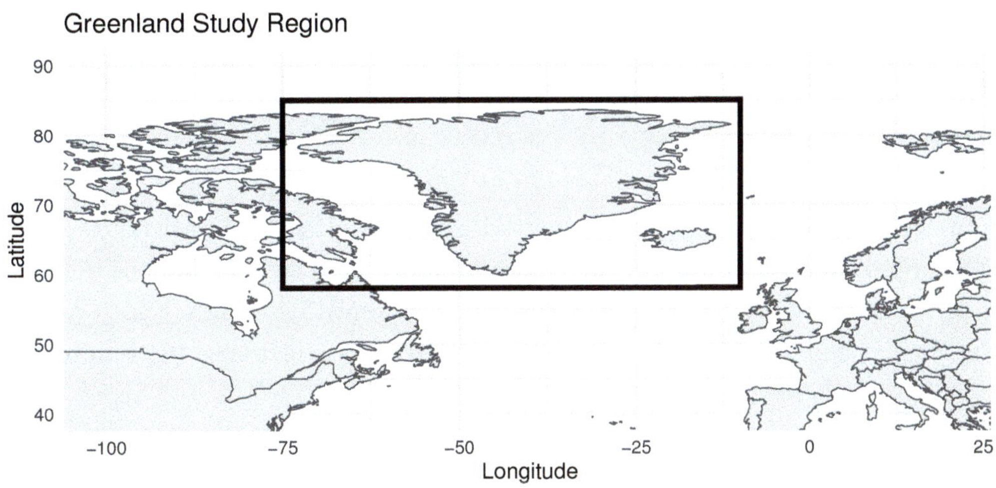

Figure 6.17: Map showing the Greenland study region used for GRACE analysis.

The black box shows the region used to compute Greenland's average thickness (Figure 6.17). This is a simple way to connect the map to the later time series.

6.7.5 Convert the map data into a time series

Now we average the grid cells inside Greenland for each month. This turns the spatial data into a single time series. A for loop is used to iterate through each time step (from 1 to `length(time)`), extract the relevant grid cells, and compute the mean thickness for Greenland at that time step. The resulting time series is stored in a data frame for easy plotting and analysis.

```
# Initialize an empty vector to store the time series
greenland_ts <- numeric(length(time))

# Loop through each time step
for (t in 1:length(time)) {
  # Extract the grid cells for Greenland at time t
  slice <- mass[lon_idx, lat_idx, t]
  # Compute the mean thickness, ignoring NA values
  greenland_ts[t] <- mean(slice, na.rm = TRUE)
}
# Create a data frame to store
```

```r
# the date and thickness time series
greenland_df <- data.frame(
  date = date,
  thickness = greenland_ts
)
```

Connecting space and time, each map becomes one number, and those numbers form a time series. This is a common workflow in geoscience: spatial data are often summarized into time series to analyze trends and patterns through time. By averaging the grid cells within the defined Greenland box, we can track how the average ice thickness changes month by month, which provides insights into the overall mass balance of the Greenland ice sheet and its response to climate change.

6.7.6 Plot the Greenland time series

The line plot shows how the ice mass changes through time. What is the first thing you notice about the plot? Are there ups and downs? This is the first step in interpreting the data. Do you notice any repeated pattern that might suggest seasonality? Is the overall trend going up or down?

```r
plot(greenland_df$date, greenland_df$thickness,
     type = "l",
     lwd = 2,
     main = "Greenland Ice Mass Change",
     xlab = "Date",
     ylab = "Equivalent Water Thickness (cm)")
grid()
```

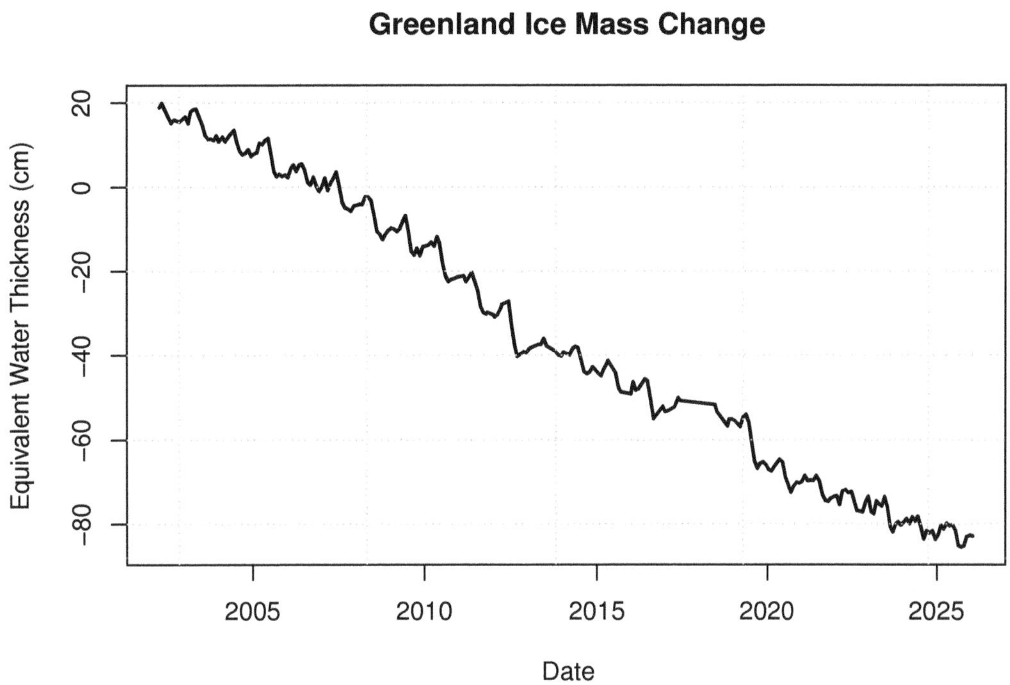

Figure 6.18: Time series of average Greenland ice mass change derived from GRACE data.

Interpretation

> The Greenland GRACE series may contain both a long-term downward trend and shorter seasonal variation. The trend is the main climate signal, while seasonality may reflect regular annual changes in snow, melt, and storage. Because GRACE data are smoothed and affected by leakage, seasonal interpretation should remain cautious. The time series lets us visualize and analyze these patterns, but remember that the values represent mass anomalies rather than precise local thickness changes. This means we can detect overall trends and seasonal patterns, but we should be cautious when interpreting exact local changes because of the coarse spatial resolution and smoothing limitations of GRACE data.

A mass anomaly means a change relative to an average reference condition.

6.7.7 Exploring Seasonality

In addition to the long-term trend, we may observe smaller repeating patterns in the time series. These patterns are called **seasonality**. We look for them using a **box plot** for each month. In a box plot, the line inside the box is the median value, the box itself covers the middle half of that month's values, and the vertical whiskers reach the rest of the typical range; any separate dots are **outliers**, single months with an unusually high or low value.

```
# Extract month information
greenland_df$month <- month(greenland_df$date, label = TRUE)

# Visualize seasonal behavior
ggplot(greenland_df, aes(x = month, y = thickness)) +
  geom_boxplot(fill = "lightblue") +
  labs(title = "Seasonal Pattern in Greenland Ice Mass",
       x = "Month", y = "Equivalent Water Thickness (cm)") +
  theme_minimal()
```

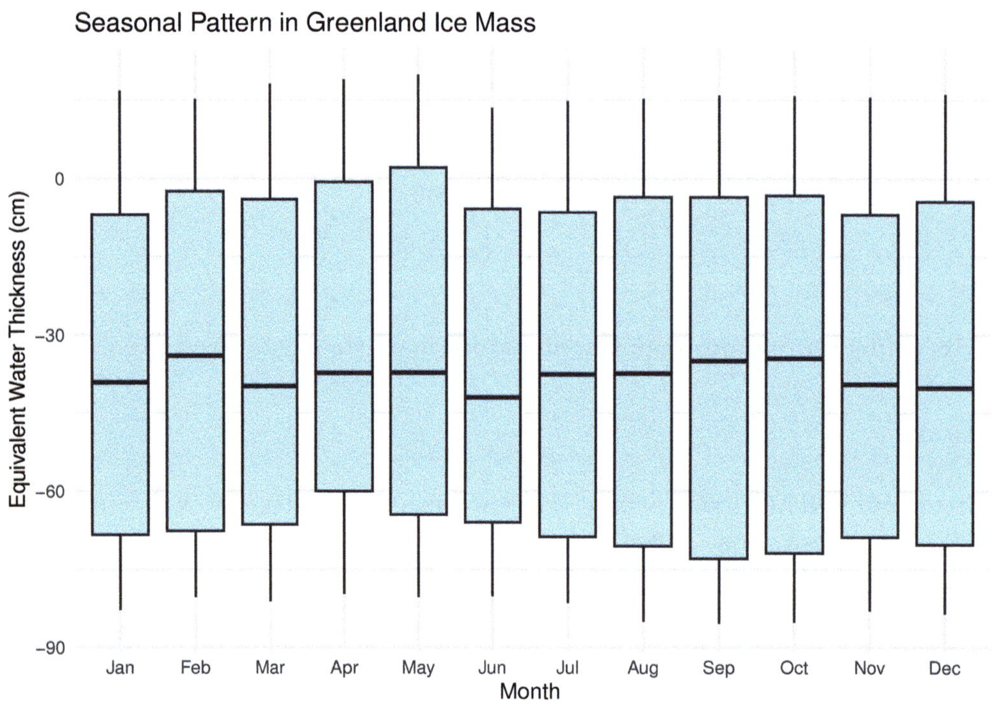

Figure 6.19: Monthly distribution of Greenland ice mass showing weak seasonal variation relative to long-term trend.

Interpretation

Median values are all negative (around -30 to -40 cm) $\rightarrow$ long-term mass loss. There is a small month-to-month structure. But the variation is weak compared to the trend. Seasonality exists, but it is relatively weak compared to the long-term decline in ice mass. In GRACE data, seasonal signals can be reduced because the data are spatially averaged over large areas and smoothed. As a result, seasonal effects appear less clearly than in higher-resolution measurements.

⚠ Think before interpreting seasonality

A weak seasonal pattern does not mean that seasonal processes are absent. It may indicate that:

- long-term trends dominate the signal
- spatial averaging reduces seasonal variability
- GRACE resolution smooths short-term changes

Focus first on the trend, then look for seasonality.

If you want a cleaner seasonal signal, subtract the trend:

```
# Estimate the trend with a 12-month moving average (each point becomes the
# average
# of the 6 months before and after it), then subtract it to expose the
# seasonal pattern.
greenland_df$trend <- stats::filter(greenland_df$thickness,
                                    rep(1/12, 12), sides = 2)

greenland_df$detrended <- greenland_df$thickness - greenland_df$trend

ggplot(greenland_df, aes(x = month, y = detrended)) +
  geom_boxplot(fill = "lightblue") +
  labs(title = "Seasonal Pattern (Detrended)",
       x = "Month", y = "Deviation (cm)") +
  theme_minimal()
```

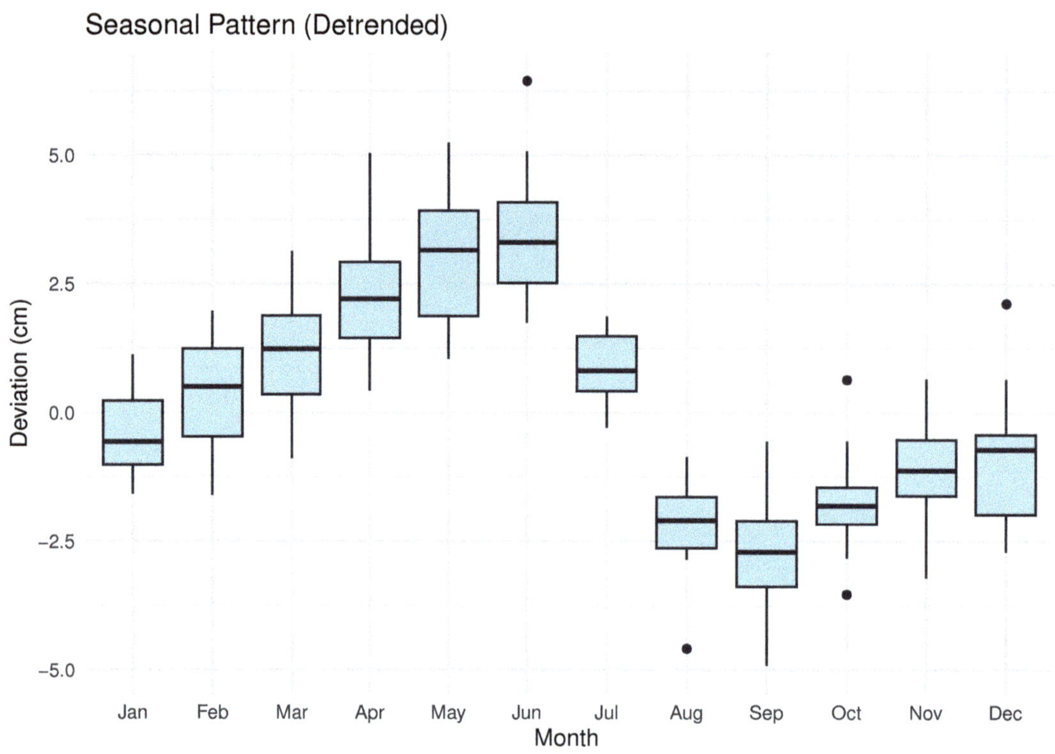

Figure 6.20: Seasonal pattern after removing the long-term trend from the Greenland ice mass time series.

Removing the long-term trend makes seasonal variation more visible (Figure 6.20).

> In real-world data, the dominant signal (trend) can mask smaller patterns (seasonality). Understanding which signal is stronger is part of scientific interpretation.

6.7.8 Compare two dates with a change map

A time series shows change through time, but a map shows **where** the change happens. Here we compare an early date and a later date. By subtracting the earlier map from the later map, we can create a difference map that shows where mass has been lost (negative values) or gained (positive values). This spatial pattern (a map made by subtracting one grid from another, pixel by pixel) can help us understand whether ice loss is concentrated in specific areas of Greenland or spread out across the entire region.

Important note about interpreting the map

The map values are not exact local thickness changes. They represent satellite-derived mass anomalies, which are influenced by smoothing and by leakage (where a strong signal in one area spreads into neighboring areas because GRACE cannot see fine detail). For that reason, the map is best used to show spatial patterns of change, while the time series is better for estimating the average annual trend. The map can show us where the most significant changes are occurring.

```
# t1 = earlier date, t2 = later date
t1 <- 20; date[t1]
```

```
[1] "2004-02-17"
```

```
t2 <- 200; date[t2]
```

```
[1] "2021-08-16"
```

```
# Difference map: later minus earlier
# Also z thickness values for t1 and t2 for later use
z_diff <- mass[, , t2] - mass[, , t1]
z_t1 <- mass[, , t1]
z_t2 <- mass[, , t2]
# Greenland bounds
lon_min <- -75
lon_max <- -10
lat_min <- 58
lat_max <- 85

# Convert longitudes from 0:360 to -180:180 for plotting
lon_plot <- ifelse(lon > 180, lon - 360, lon)

# Order longitudes so they run left to right
lon_ord <- order(lon_plot)
lon_plot <- lon_plot[lon_ord]

# Reorder the matrix to match the new longitude order
z_diff_plot <- z_diff[lon_ord, ]

# Keep only Greenland region
lon_keep <- lon_plot >= lon_min & lon_plot <= lon_max
lat_keep <- lat >= lat_min & lat <= lat_max

# Subset the data
z_greenland <- z_diff_plot[lon_keep, lat_keep]
z_t1_greenland <- z_t1[lon_keep, lat_keep]
z_t2_greenland <- z_t2[lon_keep, lat_keep]

cols <- colorRampPalette(c("#08306B", "grey78", "#FFD700"))(100)
```

```
# Use 95th percentile to avoid outliers dominating
# the color scale and to focus
# on the main spatial patterns of change
m <- quantile(abs(z_greenland), 0.95, na.rm = TRUE)

image(lon_plot[lon_keep], lat[lat_keep], z_greenland,
      col = cols,
      xlab = "Longitude",
      ylab = "Latitude",
      zlim = c(-m, m),
      main = paste("Greenland Mass Change:",
                   as.character(date[t1]), "to",
                   as.character(date[t2])))
# Add Greenland boundary as a reference
map("world",
    regions = "Greenland",
    add = TRUE,
    col = "black",
    lwd = 1.5)
```

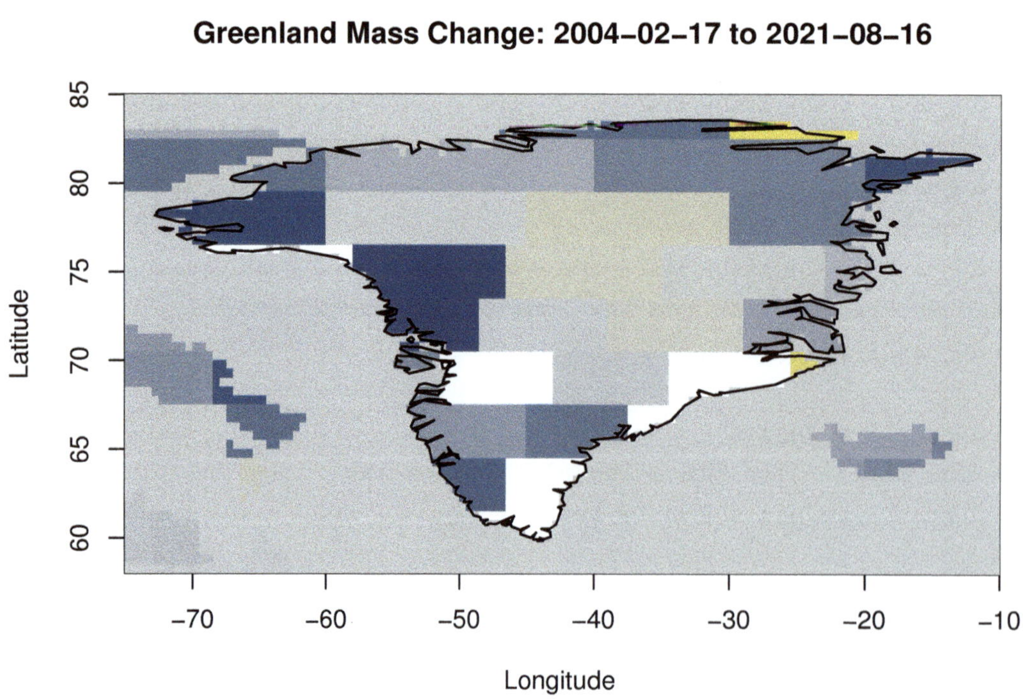

Figure 6.21: Spatial map of Greenland mass change between two selected dates derived from GRACE data (dark blue marks mass loss, gold marks mass gain).

Interpretation

- **Blue** (dark) means mass loss
- **Gold** (light) means mass gain
- **Pale grey** means little change

For a climate change story, the dark blue areas in Figure 6.21 are especially important because they can indicate ice loss.

6.7.9 Show the original map at one time step

This plot helps us see what a single GRACE map looks like before making comparisons over time. It shows the spatial pattern of mass distribution (or mass anomaly) at a given moment, rather than change. The color scale is set to highlight the range of values while avoiding extreme outliers, making the spatial patterns easier to interpret.

```
# Use 95th
zlim_t1 <- quantile(abs(z_t1_greenland), 0.95, na.rm = TRUE)
zlim_t2 <- quantile(abs(z_t2_greenland), 0.95, na.rm = TRUE)
cols <- colorRampPalette(c("#08306B", "grey78", "#FFD700"))(100)

par(mfrow=c(1,2))
image(lon_plot[lon_keep], lat[lat_keep], z_t1_greenland,
      col = cols,
      zlim = c(-zlim_t1,+zlim_t1),
      xlab = "Longitude",
      ylab = "Latitude",
      main = as.character(date[t1]))
# Add Greenland boundary as a reference
map("world",
    regions = "Greenland",
    add = TRUE,
    col = "black",
    lwd = 1.5)

image(lon_plot[lon_keep], lat[lat_keep], z_t2_greenland,
      col = cols,
      zlim = c(-zlim_t2,zlim_t2),
      xlab = "Longitude",
      ylab = "Latitude",
      main = as.character(date[t2]))
# Add Greenland boundary as a reference
map("world",
```

```
    regions = "Greenland",
    add = TRUE,
    col = "black",
    lwd = 1.5)

par(mfrow=c(1,1)) # Reset to single plot
```

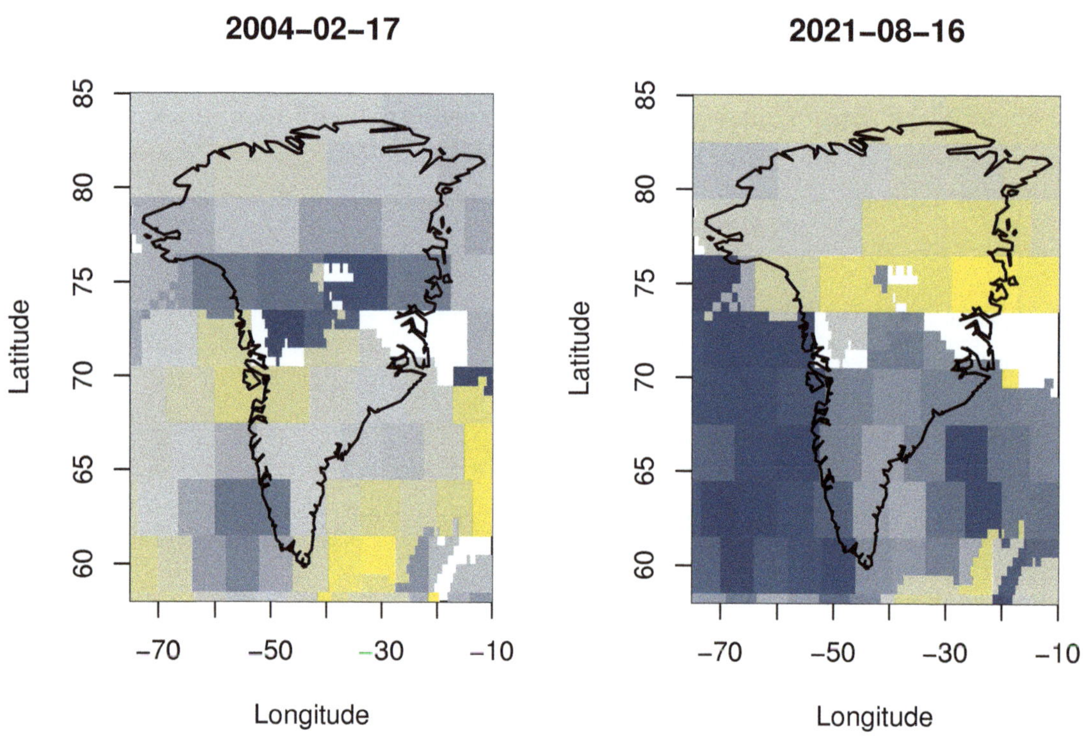

Figure 6.22: Comparison of Greenland GRACE mass patterns at two different times (dark blue low, gold high).

The color scale helps us read areas with higher and lower values (Figure 6.22). A single map shows the state at one moment; change becomes visible only when multiple maps are compared.

6.7.10 Compare the time series and the selected map together

This final figure helps connect the spatial and temporal views.

```
# par(mfrow = c(1, 2), mar = c(4, 4, 3, 5))

# Most recent time index (the snapshot shown in the map)
```

```
t <- length(time)

# Left panel: map today
z <- mass[, , t]
# Use 95th percentile to avoid outliers dominating the color scale
zlim <- quantile(abs(z_greenland), 0.95, na.rm = TRUE)
par(mfrow=c(1,2)) # Two panels side by side
fields::image.plot(
  lon, lat, z,
  col = colorRampPalette(c("#08306B", "grey78", "#FFD700"))(100),
  zlim = c(-zlim, zlim),
  xlim = c(285, 350),    # crop the view to the Greenland study region
  ylim = c(58, 85),
  xlab = "Longitude (deg E)",
  ylab = "Latitude (deg N)",
  main = paste("Greenland GRACE Map\nDate:", as.character(date[t]))
)
# Right panel: time series.
# Wide left margin (mar[2]) opens a gap from the color bar, while a small
# mgp[1]
# tucks the "Thickness (cm)" title next to THIS plot's y-axis, not the
# color bar.
par(mar = c(5, 6, 4, 2), mgp = c(2, 1, 0))
plot(greenland_df$date, greenland_df$thickness, type = "l", lwd = 2,
     xlab = "Date", ylab = "Thickness (cm)",
     main = "Greenland Thickness Over Time")
points(greenland_df$date[t], greenland_df$thickness[t],
       pch = 21, bg = "gold", col = "black", cex = 1.6, lwd = 1.5)
# Reset to single plot
par(mfrow = c(1, 1))
```

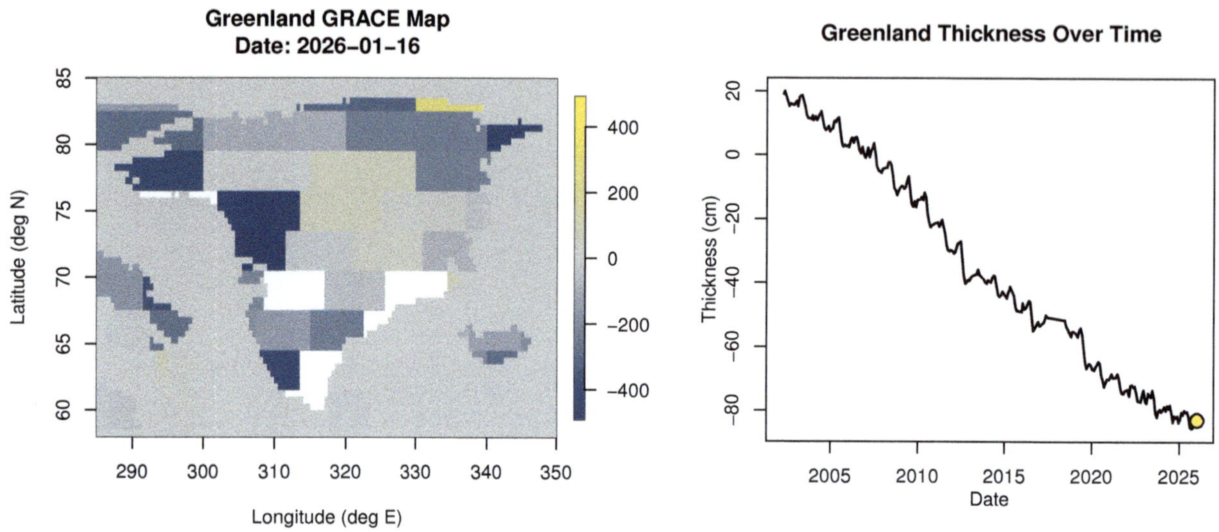

Figure 6.23: Combined visualization linking GRACE spatial map with corresponding time series thickness value.

The highlighted point (a gold dot with a black ring) on the time series marks the date shown in the map, so you can place that single snapshot within the long-term decline in ice mass (Figure 6.23).

6.7.11 What this means for climate change

This case study is not just about plotting ice data. It is about understanding a climate signal. If Greenland loses ice mass over time, that is consistent with warming conditions and contributes to sea-level rise. The exact causes are more complicated than one plot can show, but the pattern is scientifically important. This type of workflow connects Earth observation data directly to sustainability challenges related to sea-level rise, climate adaptation, and environmental resilience. GRACE data let us monitor these changes and communicate the impacts of climate change on the Greenland ice sheet. By analyzing both the time series and spatial patterns, we can gain a deeper understanding of how the ice sheet is responding to changing climate conditions and what that means for future sea-level rise.

What the time series tells us

- whether the overall ice mass is increasing or decreasing
- whether there are seasonal ups and downs
- whether the signal is stable or changing over time

What the map tells us

- where the strongest changes occur
- whether loss is spread evenly or concentrated in specific areas
- how spatial patterns support the time series trend

What the data do not tell us

- the exact physical processes causing the changes
- the precise magnitude of local thickness change due to GRACE limitations

Lessons from the Greenland Case Study

This case study shows how spatial and time series analysis can work together in a simple climate study. In it, we:

- used a GRACE NetCDF file to study ice mass change,
- defined Greenland as a region of interest,
- converted map data into a time series by averaging grid cells,
- plotted the time series to look for a trend,
- compared two dates to show where mass change occurred,
- and connected the results to climate change and sea-level rise.

The main lesson is simple: **maps show where change happens, and time series show how change happens.**

Although the three case studies examined very different Earth systems, they all drew upon concepts introduced earlier in the book. The goal was not to apply spatial analysis, time series methods, or modeling in isolation, but to combine them into complete workflows that move from raw data to scientific interpretation. The datasets changed, but the analytical principles remained the same.

> Where These Methods Lead
>
> This chapter brings the book's tools together on real Earth-system problems. One of them, the slow vertical motion of the land, is exactly what student research has measured.
> Moore and colleagues (2021) used R to study GPS records from long-running stations on Long Island, New York, asking whether the land is still rising after the last ice age or slowly sinking. After removing noise and the regular seasonal changes, they found a very slow downward motion rather than any rise, apart from a brief lift between 2013 and 2015. Even a tiny rate of sinking matters on a low, coastal island, because it adds to the effect of rising seas and a growing risk of flooding.
> The steps behind that result, cleaning a GPS series, removing its seasonal changes, and reading the long-term trend, are the same ones this book has taught you.

6.8 From Analysis to Poster

If this case study is presented on a poster, keep the message very clear and visual. You may include figures such as:

1. **Study area map:** showing Greenland and the region used for averaging.
2. **Time series plot:** showing ice mass change over time.
3. **Change map:** comparing an early date with a later date.
4. **Combined figure:** showing the selected map and the matching point on the time series.

Include these ideas in the poster text

- GRACE measures changes in Earth's mass from space.
- Greenland ice mass is an important indicator of climate change.
- A decreasing trend suggests ice loss over time.
- Spatial maps show where the strongest changes occur.
- The results should be interpreted carefully and not over-stated.

Keep the poster message short

A strong poster sentence could be:

> Satellite gravity data show a decline in Greenland ice mass over time, with the strongest changes visible in the spatial pattern of the ice sheet.

Good poster design habits

- use one main message
- keep labels large and readable
- avoid too many colors
- use a simple legend
- make sure the map and time series are easy to compare
- explain the result in plain language

⚠ Common Mistakes to Avoid

1. Reading the map as exact thickness change

The map shows mass anomalies, not precise local ice thickness. It shows patterns, not exact measurements.

2. Confusing color scale with reality

Blue and gold depend on the color scale. Always check whether the scale is centered at zero.

3. Expecting strong seasonality everywhere

Not all time series show clear seasonal cycles. In this case, the long-term trend is stronger than the seasonal signal.

4. Assuming correlation means causation
A pattern does not automatically explain why it happens. Physical interpretation must be careful.

5. Over-interpreting single maps
One map is only one moment in time. Always compare multiple times or use the time series.

6. Ignoring data limitations
GRACE data are coarse (around 300 km) and smoothed. They are best for regional trends, not local detail.

6.9 Questions to Consider Before Continuing

Before moving on, ask yourself:

- Which analytical step contributed the most useful information?
- How would the interpretation change if one dataset were unavailable?
- What additional evidence would increase confidence in the conclusions?
- Which parts of the workflow could be applied to a completely different Earth system problem?

6.10 Exercises

These exercises reuse the objects each case study already builds, so run the chapter code first. They move from quick checks to fuller analysis across the three case studies (GPS tectonics, DEM flood mapping, and GRACE ice mass).

1. Using the cleaned GPS data, plot the `east` and `north` components over time next to the `up` component you saw in the chapter. Describe in one or two sentences how the three directions differ. (Use `gps_cleaned`.)

2. Re-run the rolling mean of the GPS vertical signal with a much shorter window, such as `k = 30`, instead of 365. Compare the result with the 365-day version and explain in plain language what short-term variation reappears. (Use `rollmean(gps_cleaned$up, k = 30, fill = NA)`.)

3. The flood analysis used water levels of 0 to 30 meters. Add two intermediate levels (for example 2 and 8 meters) to the `levels` vector, rebuild the `results` table with the existing loop, and report how much of Long Island is submerged at each new level. (Use `elev` and `results`.)

4. Produce a flood map for a water level of 3 meters by adapting the chapter code that built `flood10`. Compare it with the 10-meter map and describe which areas flood first. (Use `dem_utm` and `li_v`.)

5. Choose two different time steps for the Greenland change map (for example `t1 <- 1` and `t2` near the end of the record) and recompute the difference map. Explain how the choice of dates changes the spatial pattern of mass loss you see. (Use `mass`, `lon`, `lat`, and `date`.)

6. Fit a straight line to the Greenland ice mass time series with `lm(thickness ~ as.numeric(date), data = greenland_df)`, read the slope, and convert it into an approximate change in centimeters per year. State whether the trend points to ice loss or gain, and note one reason to interpret the exact value with caution. (Use `greenland_df`.)

Getting started. For Exercise 6, you can fit and read the trend like this:

```
fit <- lm(thickness ~ as.numeric(date), data = greenland_df)
coef(fit)["as.numeric(date)"] * 365   # approximate change per year (cm)
```

A negative value means the average ice mass is decreasing over time.

The examples in this chapter focused on specific Earth systems and regional applications. However, many environmental challenges operate across much larger spatial and temporal scales.

In the next chapter, we expand from local and regional case studies to global-scale patterns related to climate, emissions, sustainability, and Earth-system change.

6.10.1 References (Data and Further Reading)

1. Blewitt, G., Hammond, W. C., & Kreemer, C. *Harnessing the GPS Data Explosion for Interdisciplinary Science.* https://geodesy.unr.edu

2. NASA Jet Propulsion Laboratory (JPL). *GRACE and GRACE-FO Mass Concentration Data.* https://grace.jpl.nasa.gov/data

3. ncdf4 package documentation. https://cran.r-project.org/package=ncdf4

4. TSA package documentation. https://cran.r-project.org/package=TSA

5. Pebesma, E. *Simple Features for R (sf package documentation).* https://r-spatial.github.io/sf

6. Hijmans, R. J. *terra: Spatial Data Analysis (R package documentation).* https://rspatial.github.io/terra

7. Wickham, H., Çetinkaya-Rundel, M., & Grolemund, G. *R for Data Science.* https://r4ds.hadley.nz

8. Lovelace, R., Nowosad, J., & Muenchow, J. *Geocomputation with R.* https://r.geocompx.org

9. Wahr, J., Swenson, S., & Velicogna, I. *Accuracy of GRACE Mass Estimates.* https://doi.org/10.1029/2005GL025305

10. Wiese, D. N., Yuan, D.-N., Boening, C., Landerer, F. W., & Watkins, M. M. (2024). *JPL GRACE and GRACE-FO Mascon Ocean, Ice, and Hydrology Equivalent Water Height CRI Filtered, Ver. RL06.3Mv04.* PO.DAAC, CA, USA. https://doi.org/10.5067/TEMSC-3JC634

11. Marsellos, A. (2026). *GRACE Mascon Data: Teaching Subset for R for Earth and Sustainability Data Analytics.* Zenodo. https://doi.org/10.5281/zenodo.20535360

12. Blewitt, G., Hammond, W. C., & Kreemer, C. (2018). *Harnessing the GPS data explosion for interdisciplinary science.* Eos, 99. https://doi.org/10.1029/2018EO104623

13. Nevada Geodetic Laboratory. (2026). *GPS time series, station NYCI, IGS20 reference frame.* University of Nevada, Reno. http://geodesy.unr.edu/gps_timeseries/IGS20/tenv3/IGS20/NYCI.tenv3

14. Moore, M., Brown, R., Thomas, M., Santella, K., Ballato, B., Marsellos, A. E., & Tsakiri, K. G. (2021). *What Can the Local GPS Network Tell Us About the Glacial Isostatic Adjustment of Long Island, NY?* In 28th Conference on the Geology of Long Island and Metropolitan New York, April 2021.

15. Parag, A. S., Marsellos, A. E., & Tsakiri, K. G. (2025). *Geostatistical Analysis of Long-Term Subsidence Effects on Isostasy of Long Island Using Continuous GPS Displacement Rates Using R.* In 32nd Conference on the Geology of Long Island and Metropolitan New York, April 2025.

16. Lawlor, J., Jones, W., Tecusan, K., & Marsellos, A. E. (2024). *Assessing Future Flood Risks: A GIS Simulation of Brooklyn's September 2023 Event with LiDAR Data.* In 31st Conference on the Geology of Long Island and Metropolitan New York, April 2024.

17. Tabarus, N., Tran, J., Achek, M., & Marsellos, A. E. (2020). *Flood Rate Investigation as a Response to Climate Change: An Example on Coney Island, New York.* In 27th Conference on the Geology of Long Island and Metropolitan New York, April 2020.

18. Foster, J. A., Marsellos, A. E., & Garver, J. I. (2011). *Predicting Trigger Level for Ice Jam Flooding of the Lower Mohawk River Using LiDAR and GIS.* In Proceedings of the 2011 Mohawk Watershed Symposium, Union College, Schenectady, NY, March 2011.

Chapter 7

Global Change and Sustainability

Visualizing Greenhouse Gas Emissions Across Countries and Time

7.1 Introduction

In the previous chapters, we learned how to work with spatial data, time series, and simple models. We used maps to understand how patterns vary across space, and time series methods to examine how systems change through time. In this chapter, we bring these ideas together to explore a global challenge: the evolution of greenhouse gas (GHG) emissions across countries and decades.

Global change refers to long-term changes in climate, ecosystems, oceans, ice, and human environments driven by both natural processes and human activity. Greenhouse gas emissions are widely studied because they connect human activities, such as energy use, transportation, and industry, to changes in climate and the environment.

Using real data from the World Development Indicators (WDI), we build a global analytical workflow that combines spatial visualization, temporal analysis, interpretation, and sustainability-focused reasoning. We produce a series of global maps showing how greenhouse gas emissions per capita evolved from 1970 to 2022. We then compare those spatial patterns with changes in total global emissions to illustrate how per-capita and aggregate measures can tell different stories.

The goal is not only to create maps but to interpret the patterns they display and what those patterns may suggest about development, inequality, industrialization, environmental pressure, and the changing relationship between human systems and the Earth system.

> **i** Main idea of this chapter
>
> In this chapter, we ask a simple question:
>
> How have greenhouse gas emissions changed across countries and through time?

To answer it, we combine:

- maps → to see where emissions are high or low
- time → to see how patterns change over decades
- interpretation → to understand what the patterns may mean

i What you will be able to do

By the end of this chapter, you will be able to:

- load and prepare a panel dataset (the same countries measured over many years) that joins map shapes to table data
- compute emissions per capita and apply a $\log_{10}$ transformation (a way of rescaling numbers) so unevenly spread values become readable on a map
- produce a multi-panel series of world maps through time using a `for` loop
- interpret spatial patterns of per-capita emissions across countries and decades
- compare per-capita maps with total global emissions to see how the two measures answer different questions
- connect the patterns to economic and policy context, and state what the data cannot explain

i Building on earlier chapters

This chapter applies tools from Chapter 3 (spatial mapping with `sf`) and Chapter 4 (reading patterns through time). If any of those ideas feel unfamiliar, it may help to revisit those sections before continuing.

7.2 Data and Variables

7.2.1 The World Development Indicators

We use publicly available data from the World Development Indicators (WDI), maintained by the World Bank. This dataset covers nearly every country in the world and includes economic, demographic, and environmental variables measured annually over several decades.

For this chapter, we use two variables:

- **Greenhouse gas emissions:** total GHG output per country, measured in megatonnes (millions of tonnes) of CO_2 equivalent per year. "CO_2 equivalent" lets different gases be added together on one scale; "AR5" refers to one standard set of conversion factors used to do that
- **Population:** total population per country, used to compute a per-capita value

> Where the indicator codes come from
>
> Each World Bank dataset has a short identifier code. You will see these codes in the data-loading block later in the chapter. You can browse the full catalog at https://data.worldbank.org or use the `WDIsearch()` function from the `WDI` package to find codes by keyword. For example, `WDIsearch("CO2")` lists every indicator related to CO_2 emissions.

These two variables have been merged with Natural Earth country boundaries (outlines of each country), so the data is laid out like a table where each row carries both a spatial geometry (the country shape) and the attribute values for one year. This is the same structure introduced in Chapter 3: the geometry tells us *where* country shapes appear on the map, and attributes tell us *what* those countries represent through data values.

7.2.2 Per Capita vs. Total Emissions

Total emissions favor large countries simply because they have more people and more economic activity. A country with one billion people will usually produce more total emissions than a country with ten million, even if individuals in the smaller country consume far more energy.

Emissions per capita divides total emissions by population. This makes countries of very different sizes directly comparable, and it shifts the question from *how much does a country emit* to *how much does each person emit on average.*

> Why use per-capita values?
>
> Per-capita values are one of the most common tools in comparative environmental analysis. They do not tell the whole story; a small wealthy country with high per-capita emissions from industry is very different from a poor country with high per-capita emissions from deforestation. They are still an essential starting point for comparison.

> A simple analogy
>
> Imagine comparing water use between a small town and a giant city. The city will almost always use more water in total because more people live there. Per-capita values instead ask

how much water each person uses on average, making places of very different sizes easier to compare fairly.

i Why We Use Per-Capita Maps

In this chapter, the maps use **emissions per capita** because they make countries of very different sizes easier to compare. Later in the chapter, we also examine **total global emissions** to show how population growth and industrial expansion affect the overall amount of greenhouse gases entering the atmosphere.
Both measures are useful, but they answer different questions.

7.2.3 The Logarithmic Transformation

Emissions per capita vary enormously across countries. Some countries emit hundreds of times more per person than others. When we map raw values, most countries appear near zero and a few outliers dominate the color scale, making the map difficult to read.

A **$\log_{10}$ transformation** solves this by compressing the range. On a $\log_{10}$ scale, each step means "ten times bigger": 10 becomes 1, 100 becomes 2, 1,000 becomes 3, and 10,000 becomes 4. So 10 and 10,000 sit far apart on a raw scale but only three steps apart here. This makes differences across the full range of countries visible at once.

! Avoiding log(0)

A tiny number written `1e-9` in R (this means 0.000000001) is added before the log transformation to avoid taking the logarithm of zero, which has no defined answer. Countries with missing or zero emissions receive a very small value that appears at the low end of the color scale.

Differences among countries may also reflect variations in reporting methods, data availability, and estimation procedures used by international agencies. These factors can introduce uncertainty and bias into the dataset, which should be considered when interpreting the maps. The log transformation helps mitigate some of these issues by reducing the influence of extreme values, but it does not eliminate underlying data quality concerns.

7.2.4 Loading the Data

The dataset used in this chapter combines two public sources: World Bank emissions and population data, and Natural Earth country boundary polygons. Both are downloaded directly through R packages; therefore no separate file downloads are needed.

The code below builds the dataset. The first time you run it, the data is downloaded directly from the World Bank's online data service (an API is just a doorway that lets programs request data automatically), cleaned, and combined with country boundaries. The final step saves the result as `world_small.rds`, so the next time you open this chapter you can load it instantly with `readRDS()` instead of waiting for the download again.

The download takes about 10 to 30 seconds, depending on your connection. Notice what is happening: in a single command, you are pulling more than fifty years of emissions and population records from nearly every country on Earth. This only needs to run once; on future runs, loading the saved file directly skips the download. There is no need to understand every line below. The comments (the lines beginning with #) explain each step in plain English, and the goal is simply to end up with one tidy dataset.

```
library(WDI)
library(sf)
library(rnaturalearth)

# Define the WDI indicator codes
indicators <- c(
  # total GHG, megatonnes CO2-eq
  ghg = "EN.GHG.ALL.MT.CE.AR5",
  # total population
  pop = "SP.POP.TOTL"
)

# Download panel data for all countries,
# all available years
panel_raw <- WDI(
  country   = "all",
  indicator = indicators,
  start     = 1960,
  end       = as.numeric(format(Sys.Date(), "%Y")),
  extra     = TRUE  # includes region, income group, ISO codes
)

# Remove summary rows (e.g. "World" or "Europe" totals) so only individual
# countries remain
panel_clean <- panel_raw[panel_raw$region != "Aggregates", ]
```

```
# Compute per capita emissions
# GHG is in megatonnes; multiply by 1e6 to convert to tonnes before dividing
panel_clean$ghg_per_capita <- (panel_clean$ghg * 1e6) / panel_clean$pop

# Load country boundary polygons from Natural Earth
world <- ne_countries(scale = "medium", returnclass = "sf")

# Join the map shapes to the data using each country's standard 3-letter
# code (e.g. USA, BRA, KEN)
world_panel <- merge(
  world,
  panel_clean,
  by.x  = "iso_a3",
  by.y  = "iso3c",
  all.x = TRUE  # keep all countries even if WDI data is missing
)

# Save the result for use in this chapter
saveRDS(world_panel, "world_small.rds")
```

This sequence of steps is a common recipe in Earth data analytics: download table data from an online service, calculate new columns from it (such as emissions per person), join it to map shapes, and save one clean, ready-to-use file. The same approach applies whether you are working with emissions, population, temperature, or any other country-level indicator.

> Reloading the dataset later
>
> On future runs, you can skip the download step entirely:
>
> ```
> countries <- readRDS("world_small.rds")
> ```
>
> This loads the saved file in less than a second, which is useful when you return to the chapter and only want to re-run the mapping code.

Now we compute the log-transformed per-capita variable and attach it to the spatial object.

```
library(sf)

# Load the preprocessed spatial panel dataset. If you have not run the
# build block above yet, run it once first to create world_small.rds.
```

```
countries <- readRDS("world_small.rds")

# Compute log10 of emissions per capita
# A small constant (1e-9) prevents log(0) errors for missing values
countries$log_ghg_pc <- log10(countries$ghg_per_capita + 1e-9)
```

After running this block, `countries` is an `sf` object (a table that also remembers each country's map shape) with one row per country per year. Each row carries the country shape and the emissions values for that year. This is called a panel dataset: the same countries are measured repeatedly across many years. The structure is the same as any other `sf` object we have worked with: geometry tells us the shape, and columns tell us the values.

7.3 Mapping Global Emissions Over Time

Now we use a `for` loop to produce one map for each time step. The loop goes through the years from 1970 to 2022 in four-year steps, selects the data for that one year, and draws a map. This is the same loop structure introduced in Chapter 1. Here it does more: it replaces what would be fourteen separate manual steps in a graphical interface with a single automated workflow.

```
# arrange panels in a 5-row by 3-column grid
par(mfrow = c(5, 3))
# tight margins so maps fill the space
par(mar = c(0, 0, 6, 0))

for (yr in seq(1970, 2022, by = 4)) {

  # Subset to the current year
  map_data <- countries[countries$year == yr, ]
  # Get the values and split them into tidy number ranges, one color per range
  vals     <- map_data$log_ghg_pc
  breaks   <- pretty(vals)

  # Draw the country shapes, coloring each one by its log GHG per capita
  # hcl.colors(..., "YlOrRd", rev = TRUE) runs from light yellow (low)
  # to dark red (high); cut() decides which color band each country
  # falls into. (You do not need to memorize this line; the comments
  # explain the idea.)
  plot(
    st_geometry(map_data),
    col    = hcl.colors(length(breaks) - 1, "YlOrRd",
```

```
                            rev = TRUE)[cut(vals, breaks)],
    border = NA
  )

  title(main = yr, line = -0.7, cex.main = 2)
}

# Final panel: shared legend
plot.new()
legend(
  "center",
  legend = c("High", "Low"),
  fill   = hcl.colors(20, "YlOrRd", rev = TRUE)[c(20, 1)],
  title  = "log(GHG per capita)",
  bty    = "n",
  cex       = 2,    # label size
  pt.cex    = 2,    # color box size
  title.cex = 2     # title size
)
```

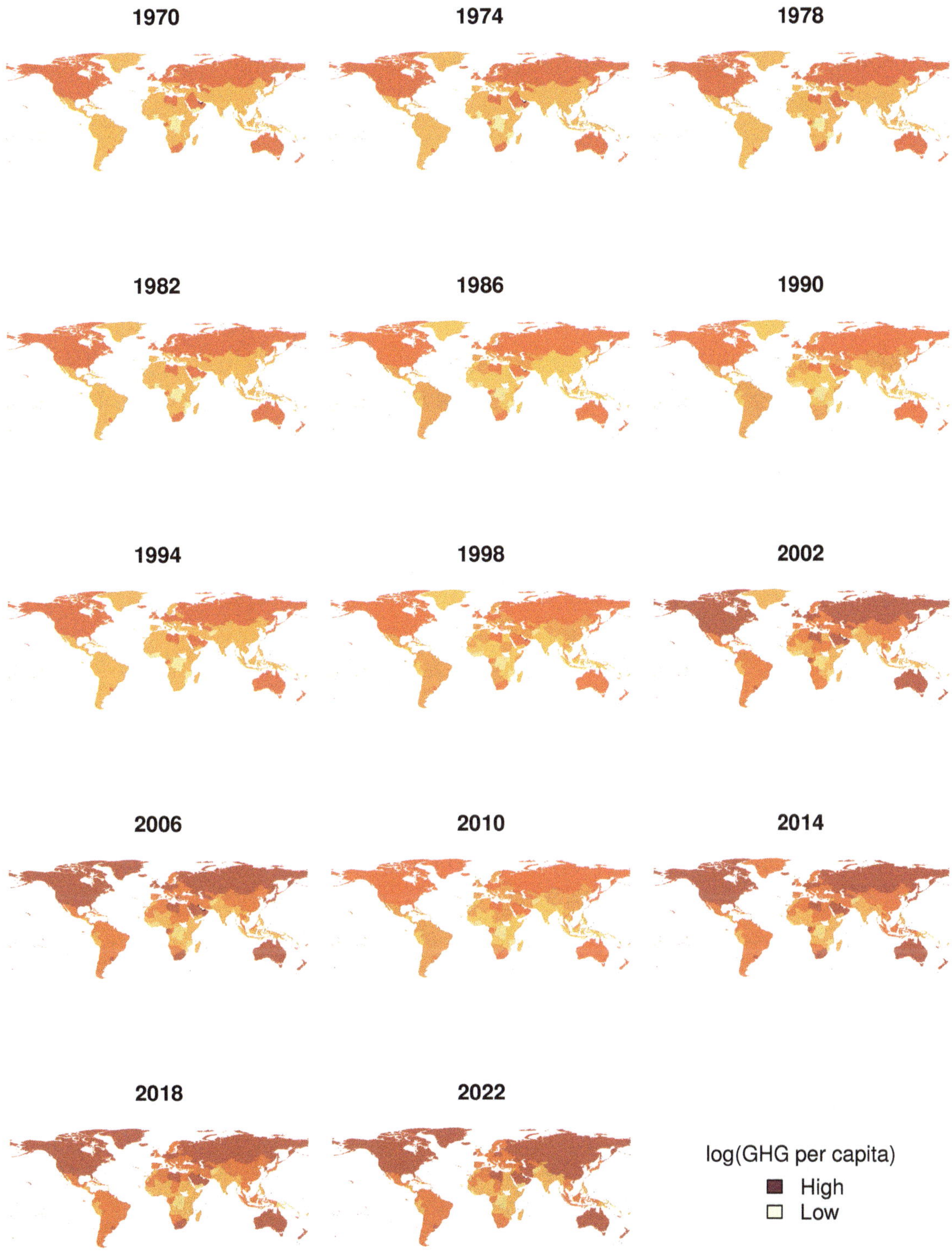

Figure 7.1: Log-transformed GHG emissions per capita across countries, shown at 4-year intervals from 1970 to 2022. Dark red indicates higher emissions per capita; light yellow indicates lower. Each panel uses its own color scale, so the maps show relative differences within each year.

Each panel in Figure 7.1 corresponds to one year. Even before detailed interpretation, broad regional differences become immediately visible across the maps. Reading left to right and top to bottom moves forward through time from 1970 to 2022. The colors run continuously from light yellow through orange to dark red, so an orange country sits at a middle level, between the highest (dark red) and lowest (light yellow).

i How to read these maps

- Each panel shows one year. Reading left to right, top to bottom, moves forward in time from 1970 to 2022
- **Dark red** countries have higher log-transformed GHG emissions per capita
- **Light yellow** countries have lower log-transformed GHG emissions per capita
- Each panel is scaled independently, so the maps show *relative differences within a year*, not absolute comparisons across years

This means that a country shown in the same color in two different years may not represent the same exact emissions value in both panels.

7.3.1 What the Maps Tell Us

- Developed countries in North America, Europe, and Australia consistently appear in darker red colors across all time periods, reflecting persistently high per-capita emissions.
- Many countries in sub-Saharan Africa and South Asia remain in lighter yellow colors throughout, indicating lower per-capita emissions over the full period.
- From the 1990s onward, parts of East Asia and the Middle East gradually shift toward darker colors. This reflects increasing industrial activity and rising energy consumption in those regions.
- In the most recent panels, some traditionally high-emission countries show slightly lighter colors, consistent with energy transitions and efficiency improvements.

7.3.2 What the Maps Do Not Tell Us

- The maps show per-capita values, not total national output. A small wealthy country can appear dark red even if its total contribution to global emissions is minor.
- The maps show *that* patterns change, not *why*. Interpretation requires additional context, which we address in the following sections.
- Student research using the same World Development Indicators data reached a similar limit. Burgos et al. (2026) found that even the strongest single predictor of emissions, a country's GDP per person, still left most of the differences between countries unexplained, a reminder

that maps and models reveal patterns but cannot, on their own, explain why they occur or decide what counts as sustainable.

- Countries with no data for a given year are left uncolored (shown in plain gray or white, not light yellow) and should not be read as low emissions.

> ⚠ Colors are not directly comparable across years
>
> Avoid reading these maps as a direct comparison of emission levels across time. Because each panel is scaled independently, a country appearing the same color in 1970 and in 2010 may have very different absolute emissions in each year.

> ℹ Quick Checkpoint
>
> Before continuing, make sure you understand:
>
> - why we use emissions per capita instead of only total emissions,
> - why the maps use a log transformation,
> - and why colors cannot be directly compared across years when each panel uses its own scale.

7.4 Interpreting the Patterns

7.4.1 Developed vs. Developing Countries

The most consistent pattern across all panels is the contrast between industrialized and non-industrialized regions. North America, Western Europe, and Australia show persistently high per-capita emissions. That pattern reflects decades of heavy reliance on fossil fuels for electricity, transportation, and industrial development throughout the 20th century.

Countries in sub-Saharan Africa and parts of South and Southeast Asia maintain lower per-capita values throughout the period. That does not mean those regions are unaffected by global change; it means that each individual person contributes less to the global total, often because access to energy itself is more limited.

7.4.2 Post-2000 Shifts

Around the early 2000s, a visible shift begins in parts of East Asia and the Persian Gulf region. Those years saw rapid industrial expansion in China, India, and several oil-exporting economies. Per-capita emissions in those regions rise noticeably across successive panels. At the same time, some Western European countries begin to show slightly lighter colors in the most recent panels. The trend is consistent with the rollout of renewable energy capacity and efficiency legislation in the EU during the 2000s and 2010s.

> **i** Four-year snapshots simplify reality
>
> These shifts do not happen overnight. They are gradual, and the four-year interval between panels means that rapid transitions may appear smoother than they actually were.

7.4.3 Regional Differences

No two regions follow the same path. The Middle East shows high and rising per-capita emissions driven by energy-intensive economies built around petroleum production and low domestic energy prices. Latin America shows moderate values with slower growth. Much of Africa remains at lower per-capita levels throughout, not because emissions have fallen but because industrial development has remained limited relative to population growth.

Interpreting these maps requires thinking about several factors at the same time. Population size, industrial structure, energy access, policy choices, and trade patterns all shape what we see. A single color on a map rarely has only one explanation.

7.4.4 Two Different Views of the Same Problem

The multi-panel maps above show per-capita values, and in many regions those values appear to stabilize or even decline in recent decades. This might suggest that global emissions are leveling off. However, the global total emissions trend tells a different story.

The plot below shifts perspective and examines **total greenhouse gas emissions worldwide**. This allows us to compare individual-level patterns with the overall amount of greenhouse gases entering the atmosphere.

```
par(mar = c(5, 5, 5, 4))
# restore panels to a 1-row by 1-column grid
par(mfrow = c(1, 1))
# Add together every country's emissions within each year to get the world
# total
# (the ~ below simply reads as "ghg grouped by year")
totals <- aggregate(
  ghg ~ year,
  data = countries,
  sum,
```

```
  na.rm = TRUE
)

plot(
  totals$year[totals$year <= 2022],
  totals$ghg[totals$year <= 2022],
  type = "l",
  xlab = "Year",
  ylab = "Total GHG emissions (Mt CO2-eq)",
  main = "Global total GHG emissions"
)
grid()
```

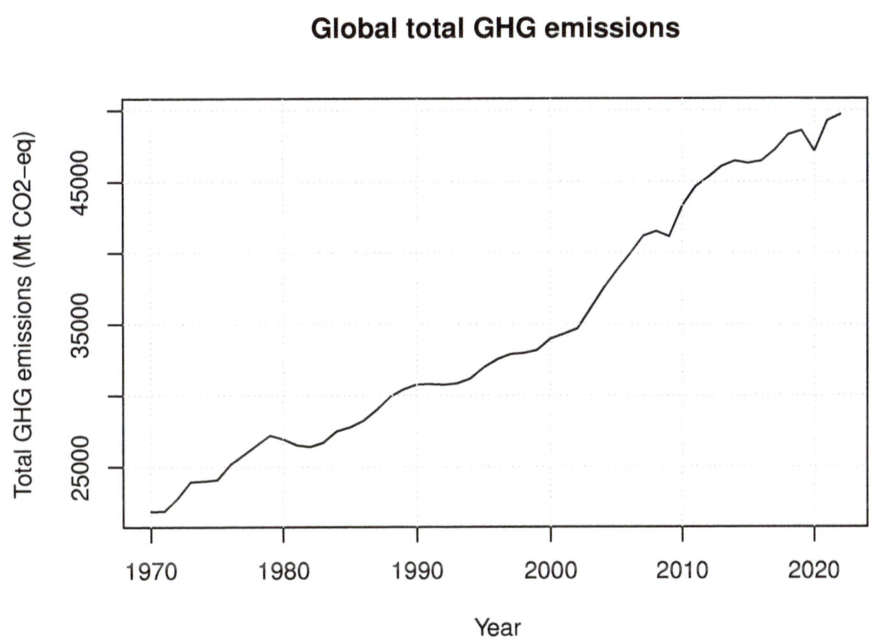

Figure 7.2: Total greenhouse gas emissions across all countries by year. Even as per-capita emissions stabilize or decline in some regions, total global emissions can continue to rise because of population growth and industrial expansion.

> **i Reading the global total**
>
> This figure shows the sum of greenhouse gas emissions from all countries combined. Unlike the maps shown earlier, which display emissions per capita, this plot focuses on the total amount of greenhouse gases entering the atmosphere each year.

The global sum continues to rise even as per-capita values in some countries appear flat or declining. Three forces are working in the same direction. First, population growth adds more people whose emissions count toward the total. Second, industrialization in previously low-emission countries adds new emitters to the global picture. Third, some apparent declines in high-income countries are partly caused by manufacturing moving to other countries. When factories move, the emissions move with them, even if the products are still consumed in the countries that now import them. A country can show a falling per-capita trend while the goods it imports carry an increasing emissions footprint produced elsewhere.

> **i** Two measures, two questions
>
> This distinction between per-capita and aggregate trends is one of the most common sources of confusion in public discussions about climate data. Both measures are correct; they answer different questions. Per-capita values measure individual-level intensity. The global sum measures the total burden on the atmosphere. Both matter, and they do not always point in the same direction.

7.5 Policy and Economic Context

7.5.1 Outsourcing of Emissions

One important limitation of counting emissions country by country is that it records *where* goods are produced, not *where* they are bought and used. A country that outsources its manufacturing will show lower domestic emissions, even if its consumers are responsible for driving that production elsewhere. This means that a decline in emissions in one country may correspond to a rise in another. The global total is what ultimately matters for the climate, but the distribution across countries is shaped by trade as much as by domestic choices.

> **i** Production vs. consumption emissions
>
> Consumption-based emissions accounting assigns emissions to the consumer rather than the producer. This often tells a different story than the production-based data we use here. Both perspectives are valid, and comparing them is a productive research question in environmental economics.

7.5.2 Industrial Growth and Energy Demand

Countries that industrialize rapidly tend to increase emissions quickly, because early industrial infrastructure is almost always fossil-fuel-dependent. Building factories, roads, power plants, and cities requires enormous amounts of energy, and the cheapest available energy has historically been coal, oil, or gas. This is the pattern visible in parts of East Asia and South Asia in the post-2000 panels. It reflects the same developmental trajectory that industrialized nations followed a century earlier, under conditions where fossil fuels were the dominant available option.

7.5.3 Energy Transition

The gradual shift toward lighter colors visible in some European countries in recent panels is consistent with real policy action: carbon pricing, renewable energy subsidies, building efficiency standards, and in some cases industrial restructuring. These transitions take decades and are often uneven within countries, but they are detectable in national-level data.

> Data show what, not why
>
> The data cannot tell us whether a country's emissions declined because of policy, economic slowdown, outsourcing, or some combination of all three. That is precisely why data-driven analysis must be paired with contextual knowledge. Patterns reveal *what* happened. Understanding *why* requires more.

7.6 R vs. Graphical Software

This chapter produced fourteen maps using a loop of fewer than twenty lines. In a graphical interface, the same result would require loading the data, filtering to each year, adjusting the color scale, and exporting the figure fourteen times, by hand, with no guarantee that each step was performed identically.

> Why Reproducibility Matters
>
> When every step is written in code, another researcher can re-run the exact same analysis, verify the results, and build directly on the work. This is a core requirement of open and transparent science.

The loop structure used here is the same one introduced in Chapter 1 and applied throughout the book. By Chapter 7, you have seen it used for vectors, GPS time series, flood thresholds, and now global emissions maps. The syntax stays the same. Only the data and the question change. This consistency is what makes coding a transferable skill rather than a collection of isolated manual steps.

7.7 Chapter Summary

This chapter brought together the major ideas developed throughout the book. Spatial analysis helped us examine where greenhouse gas emissions occur, time-based analysis helped us examine how those patterns change, and sustainability concepts helped us interpret why those changes matter. Starting from raw WDI data, we computed per-capita emissions, applied a log transformation to make the distribution readable, and produced a series of fourteen maps using a single loop. We then compared those patterns with changes in total global emissions through time.

The maps displayed consistent patterns: persistently high per-capita emissions in industrialized regions, gradual increases in rapidly developing economies after 2000, and early signs of stabilization in some high-income countries. We connected those patterns to three explanatory factors: trade and outsourcing, industrial development, and energy transition, while noting that data alone cannot distinguish between them.

The workflow demonstrated a principle that runs through this entire book: coding allows you to define an analytical process once and apply it automatically across many cases. What would have required **fourteen manual repetitions** in a graphical tool was accomplished in a single reproducible script. More importantly, the same workflow can be applied to entirely different datasets. Once the analytical process has been defined, only the data and the scientific question need to change.

7.8 Final Reflection

Earth data analytics is not about producing maps. It is about asking what the data and maps mean. In this chapter, we used the tools built across the previous six chapters, including spatial objects, loops, log transformations, multi-panel figures, and applied them to a question that connects data to the real world: how are human emissions distributed across countries, and how have they changed over more than fifty years?

Understanding global change requires more than a color scale. It requires recognizing what per-capita values and total emissions each capture, and how the two perspectives can lead to different interpretations. It also means seeing what trade flows conceal, and what policy signals might appear in a gradual color shift across panels. The data reveal patterns. Interpretation connects those patterns to real Earth systems and human activities. This is the core of Earth Data Analytics and Sustainability: using data as a lens to understand systems, not as a substitute for thinking about them.

As you reflect on the chapter, consider:

- Which conclusions come directly from the data and which require interpretation?
- How do maps and time series complement one another when studying environmental change?
- What advantages did automation provide when producing many maps through time?
- Which factors beyond geography help explain the observed emissions patterns?

7.9 Lessons from Mapping Change Through Time

This chapter combined spatial analysis, temporal analysis, and sustainability-focused interpretation to examine how greenhouse gas emissions vary across countries and decades. Using a single automated workflow, we transformed a series of individual maps into a spatial record of change through time.

- Greenhouse gas emissions per capita allow fair comparison across countries of very different sizes.
- Total emissions provide a different perspective by showing the overall amount of greenhouse gases entering the atmosphere.
- A $\log_{10}$ transformation compresses wide-ranging data so that patterns across the full distribution become visible in a single color scale.
- A `for` loop can automatically generate an entire sequence of maps through time, making the workflow reproducible, scalable, and easy to update when new data become available.
- Developed regions show persistently high per-capita emissions; rapidly industrializing regions show the strongest growth after 2000.
- Emissions patterns reflect trade, development stage, and energy policy, not just geography. Data show *what*; interpretation requires understanding *why*.
- Maps show *where* change happens; time series show *how* it unfolds over time. Both perspectives together give a more complete picture.

Where These Methods Lead

This final chapter looks at global sustainability, where data and models meet questions that numbers alone cannot answer. Student research reached the same limit.

Burgos and colleagues (2026) used R to model the greenhouse-gas emissions of 148 countries from World Bank development data. Of every measure they tried, a country's wealth per person was the strongest single factor, yet the model still explained only about a third of the differences between countries. Most of the story lay elsewhere, in policy, energy systems, and choices that no single number captures. Their conclusion matches this chapter's: a model can reveal a pattern, but people, not the model, must decide what it means and what counts as sustainable.

Seeing clearly what a model can and cannot say is what this book has tried to teach.

7.10 From Analysis to Poster

When you present a sustainability analysis on a poster or slide, keep it clear and honest about its limits:

- open with the question and why it matters (for example, what shapes a country's emissions)
- show one clear map or chart, not many, and let the strongest pattern stand out
- state the main finding in a single plain sentence
- give the size of the effect and how much it leaves unexplained, so the audience sees the limits
- separate what the data shows from what it cannot say about cause, or about what counts as fair
- end with the human question: what should we do, and who decides

A strong sustainability poster shows the pattern clearly and is honest that what to do about it is a human choice, not one the numbers can make.

7.11 Exercises

The following exercises use the same `world_small.rds` dataset.

Exercise 1: Single-year map Produce a map of log GHG per capita for the year 2010 only. Label it clearly and add a legend. Which regions stand out as highest and lowest?

Exercise 2: Subset by region Filter the dataset to countries in one world region of your choice (use the `region` column from the WDI data). Reproduce the multi-panel map series for that region only. How do patterns within the region compare to the global picture?

Exercise 3: Raw vs. log scale Produce two maps for the same year: one using raw `ghg_per_capita` values and one using `log_ghg_pc`. Compare them side by side using `par(mfrow = c(1, 2))`. In your own words, explain why the log transformation changes what is visible.

Exercise 4: Time series for one country Choose one country and extract its emissions per capita values across all available years. Plot these as a line graph using `plot(..., type = "l")`. Does the trend match what you expected from the maps? Describe the pattern in two or three sentences.

> Getting started
>
> For Exercise 4, you can extract one country's data using:

```r
country_data <- countries[countries$name == "Brazil", ]
plot(country_data$year, country_data$ghg_per_capita,
     type = "l", main = "Brazil GHG per capita",
     xlab = "Year", ylab = "Tonnes CO2-eq per person")
```

Replace `"Brazil"` with any country name present in the dataset.

7.11.1 References (Data and Further Reading)

1. World Bank. *World Development Indicators (WDI).* https://databank.worldbank.org/source/world-development-indicators
2. Natural Earth. *Natural Earth Map Data.* https://www.naturalearthdata.com
3. Pebesma, E. *Simple Features for R (sf package documentation).* https://r-spatial.github.io/sf
4. Lovelace, R., Nowosad, J., & Muenchow, J. *Geocomputation with R.* https://r.geocompx.org
5. Wickham, H., Çetinkaya-Rundel, M., & Grolemund, G. *R for Data Science.* https://r4ds.hadley.nz
6. Intergovernmental Panel on Climate Change (IPCC). *Climate Change 2023: Synthesis Report.* https://www.ipcc.ch/report/ar6/syr
7. Our World in Data. *CO_2 and Greenhouse Gas Emissions.* https://ourworldindata.org/co2-and-greenhouse-gas-emissions
8. United Nations Environment Programme (UNEP). *Emissions Gap Report.* https://www.unep.org/resources/emissions-gap-report
9. NASA Earth Observatory. *Global Climate Change and Earth System Observations.* https://earthobservatory.nasa.gov
10. Natural Earth. (n.d.). *Free vector and raster map data at 1:10m, 1:50m, and 1:110m scales.* https://www.naturalearthdata.com
11. South, A. (2017). *rnaturalearth: World Map Data from Natural Earth [R package].* https://CRAN.R-project.org/package=rnaturalearth
12. World Bank. (2026). *World Development Indicators.* The World Bank Group. https://databank.worldbank.org/source/world-development-indicators
13. Burgos, J., Desilva, A., Wagner, J., Marsellos, A. E., & Yetgin, E. (2026). *AI-Driven Analysis of Global Emissions: Insights and Ethical Limits for Sustainable Decision-Making.* 13th Annual NJBDA Symposium, New Jersey Big Data Alliance, Glassboro, NJ, May 2026.

Glossary

This glossary provides **selected key terms** used throughout the book. It is not intended to be exhaustive, but to offer quick reminders of important concepts introduced in context within the chapters.

A

ADF test (Augmented Dickey-Fuller test): A statistical test that checks whether a time series is stationary. A small p-value (below 0.05) is evidence that the series is stationary.

argument: Information supplied to a function. Arguments control how a function behaves and what data it uses.

array: A data structure that stores values in multiple dimensions, such as rows, columns, and layers.

artificial intelligence (AI): Computer systems designed to perform tasks that normally require human intelligence, such as pattern recognition, prediction, and decision support.

aspect (terrain): The compass direction that a slope faces. It is used together with slope to create a hillshade.

attribute: Descriptive information associated with a spatial feature or dataset, such as a name, population, elevation, or category.

attribute table: The tabular information attached to spatial features. Each row corresponds to a spatial object and each column stores descriptive information.

autocorrelation: Measures how a time series relates to its past values. It helps identify memory and repeating patterns in the data.

B

bathymetry: The measurement and mapping of underwater terrain and seafloor depth.

box plot: A compact summary of a group of values: the line inside the box is the median, the box spans the middle half of the values (the first to third quartile), the whiskers reach the rest of the typical range, and separate dots mark outliers.

C

clipping: A spatial operation that keeps only the portion of a dataset that falls within a specified boundary.

coefficient: The number a model assigns to a predictor. With a single predictor it is the slope, telling you how much the response changes for a one-unit change in that predictor.

contrast stretch: A visualization adjustment that spreads pixel values across the full brightness range, similar to a brightness and contrast control in a photo editor, so faint differences become easier to see.

coordinate reference system (CRS): Defines how locations on Earth are represented and how coordinates relate to real-world positions. All spatial layers must use the same coordinate reference system to be compared, aligned, and analyzed correctly.

correlation: A measure of the strength and direction of a relationship between two variables.

cross-correlation: Compares two time series to determine whether they move together and whether one tends to lead or lag the other.

D

data frame: A table of data with rows and columns. Each column represents a variable, and each row represents an observation.

data type: The kind of information stored in an object, such as numeric values, text, dates, or logical values.

decision tree: A model that makes predictions by following a branching set of yes/no questions, with each answer sending the data down a different path until it reaches an estimate.

decomposition: Splitting a time series into separate parts, usually a long-term trend, a repeating seasonal cycle, and the remaining short-term variation.

DEM (Digital Elevation Model): A raster dataset where each cell represents elevation. It is used to study terrain and surface processes.

deseasonalizing: Removing the repeating seasonal cycle from a time series so that other patterns become easier to see.

detrending: Removing the overall upward or downward drift from a time series so that the remaining variation can be studied on its own.

duplicate record: A repeated observation that appears more than once in a dataset.

E

elevation: The height of a location relative to a reference level, usually mean sea level.

EPSG code: A short ID number that identifies a specific coordinate system. For example, 4269 is NAD83, 4326 is WGS84, and 32618 is UTM Zone 18N.

equivalent water thickness (EWT): A way of expressing a change in mass as the thickness of a layer of water, in centimeters, that would produce the same change. It is commonly used with GRACE satellite gravity data.

extrapolation: Estimating values beyond the range of the observed data. It is less reliable than interpolation because no nearby measurements are available to guide it.

F

frequency: The rate at which a repeating pattern occurs within a time series.

function: A set of instructions that performs a task. Functions take inputs (arguments) and return an output.

G

geometry: The spatial shape and location of a feature, represented as a point, line, or polygon. Geometry is stored together with attributes in a spatial dataset.

GeoPackage: A modern spatial file format that stores vector and raster data in a single file.

GIS (Geographic Information System): A system used to store, analyze, visualize, and interpret spatial data linked to geographic locations.

greenhouse gas (GHG): A gas that traps heat in the atmosphere, including carbon dioxide, methane, and nitrous oxide.

H

hazard: A natural or human-induced process that has the potential to cause harm to people, property, or the environment.

hillshade: A visualization technique that simulates sunlight on a terrain surface to enhance topographic features.

I

indexing: The process of selecting specific elements, rows, columns, or positions within an object.

intercept: In a fitted line, the predicted value of the response when the predictor is zero. It sets where the line crosses the vertical axis.

interpolation: A method used to estimate missing values between surrounding known values. It assumes that change between points is smooth.

L

lag: A time shift between observations or between two time series.

leakage: In GRACE satellite data, the unwanted spreading of a signal from one area into neighboring areas, caused by the limited spatial resolution.

library: A collection of functions and tools loaded into R through a package.

line (spatial): A vector feature representing a connected path, such as a road, river, coastline, or fault.

linear regression: A method that fits a straight line through data to describe how one variable changes with another.

list: A flexible data structure that can store different types of objects together, including vectors, data frames, or other lists.

loop: A structure that repeats a set of commands multiple times. Useful for automating repetitive tasks.

M

machine learning: A branch of artificial intelligence that uses algorithms to learn patterns from data and make predictions.

mass anomaly: A difference in mass compared with a long-term average, such as the gain or loss of ice or water in a region.

mean absolute error (MAE): The average size of the prediction errors, ignoring their direction. Smaller values indicate better predictions.

merge: The process of combining datasets using one or more common variables.

missing values: Values that are not available in the dataset. They must be identified and handled before analysis.

model: A simplified representation of reality used to describe patterns or make predictions. Models help us understand relationships in data.

Modified Julian Day (MJD): A continuous day count used to label dates as single numbers, measured from November 17, 1858.

moving average: A smoothing method that replaces each value with the average of nearby values inside a sliding window, reducing short-term noise so broader patterns stand out.

N

NAD83: A geographic coordinate system, based on latitude and longitude, commonly used for locating features in North America.

NetCDF: A file format for storing multi-dimensional scientific data, such as values that vary across longitude, latitude, and time.

NGVD 1929: The National Geodetic Vertical Datum of 1929, a fixed national reference level first set from average sea level. Heights such as groundwater levels are reported as feet above this reference, where 0 is roughly average sea level.

noise: Random variation in data that does not follow a clear pattern. Noise is always present in real-world datasets.

normalization: Rescaling data to a common range, often between 0 and 1, so that variables can be compared more easily.

O

object: A named container used to store data in R. Objects can hold numbers, text, vectors, or more complex structures.

outlier: A value that lies far from the rest of the data and may reflect an error or an unusual event.

overfitting: A modeling problem where a model memorizes the random variations of one dataset instead of the real pattern, so it works poorly on new data.

P

package: A collection of functions, datasets, and tools that extend the capabilities of R.

per capita: A quantity expressed per person within a population.

periodicity: The tendency of a process or pattern to repeat at regular intervals through time.

periodogram: A method used to identify repeating cycles in a time series by checking which cycle lengths (how often a pattern repeats) are strongest in the data.

pixel: A single cell in a raster dataset. Each pixel stores one value, such as elevation, temperature, or rainfall, for a specific location.

point (spatial): A vector feature representing a single geographic location.

polygon: A vector feature representing an enclosed area such as a county, state, or lake.

prediction: An estimate of future or unknown values based on a model.

predictor: An input variable that a model uses to estimate the response.

projection: A mathematical transformation that converts Earth's curved surface into a flat map representation so that spatial data can be displayed, measured, and analyzed.

p-value: A number between 0 and 1 that measures how surprising a result would be if there were no real effect. Small values, often below 0.05, suggest stronger evidence.

Q

quartile: A value that divides sorted data into four equal parts. The first quartile (1st Qu.) sits a quarter of the way through the data, and the third quartile (3rd Qu.) sits three quarters of the way through.

R

random forest: A model that builds many decision trees and averages their results to improve prediction.

random walk: A time series whose next value is a random step away from the last one, so it drifts with no fixed level to return to. It is a non-stationary pattern.

raster data: A grid-based data structure where each cell contains a value, such as elevation, temperature, or rainfall.

raster resolution: The size of individual cells within a raster dataset. Smaller cells provide more detail.

region of interest (ROI): A defined geographic area selected for analysis.

regression: A statistical method used to model and quantify relationships between variables.

reproducibility: The ability to repeat the same analytical workflow and obtain the same results using the same data and code.

residual: The difference between an observed value and the value predicted by a model.

response: The output variable that a model tries to predict.

root mean square error (RMSE): A measure of prediction error that penalizes large mistakes more than small ones. It is expressed in the same units as the predicted variable.

R-squared: The proportion of the variation in the response that the model explains, ranging from 0 (none) to 1 (all).

S

script: A saved collection of R commands that can be run and reused.

seasonal-trend decomposition (STL): A method that separates a time series into trend, seasonal, and remainder components.

seasonality: A repeating pattern in a time series that occurs at regular intervals, such as yearly or monthly cycles.

shapefile: A commonly used GIS file format for storing vector spatial data and associated attributes.

signal: The meaningful pattern contained within data, distinct from random noise.

slope: In a fitted line, how much the response changes for each one-unit increase in the predictor.

slope (terrain): How steep the land is at a location. It is used together with aspect to create a hillshade.

smoothing: A technique used to reduce short-term fluctuations in data in order to highlight long-term patterns.

spatial analysis: The study of patterns, relationships, and processes that involve geographic location.

spatial layer: A dataset containing geographic information that can be displayed, combined, and analyzed on a map.

spectral analysis: A method used to identify repeating cycles within a time series and which of those cycles are the strongest.

spectral density: A measure of how strong each repeating cycle is in a time series. Tall values in a periodogram mark the dominant cycles.

standard deviation: A measure of how spread out values are around their average.

stationarity: A property of a time series where the statistical behavior (such as mean and variance) remains relatively stable over time.

statistical significance: An indication that a result is unlikely to be due to chance alone.

study area: The geographic region selected for investigation and analysis.

T

testing data: The subset of data used to evaluate model performance.

time series: A sequence of observations collected through time, usually at regular intervals.

training data: The subset of data used to build a model.

trend: The long-term direction of a time series, showing whether values increase, decrease, or remain stable over time.

U

UTM (Universal Transverse Mercator): A projected coordinate system that divides Earth into zones and measures location in meters.

V

validation: The process of checking whether data or model results are reasonable, accurate, and suitable for analysis.

variance: A measure of how spread out values are around their average. It is the square of the standard deviation.

vector: A simple list of values that are all the same kind (for example, all numbers or all words). Vectors are one of the most common ways to store data in R.

vector data (spatial): Spatial data represented as points, lines, or polygons. Used to describe discrete features such as locations and boundaries.

vertical exaggeration: Temporarily multiplying elevation values so that small differences in flat terrain become easier to see in a map or hillshade.

W

WGS84: A widely used geographic coordinate system based on latitude and longitude that serves as the foundation for GPS positioning.

Z

z-score: How many standard deviations a value lies from the average. Large z-scores, for example beyond 3, flag unusual values or outliers.

Final Note

This glossary is intended as a quick reference to support your learning. If a term feels unclear, you may find it helpful to revisit the chapter where it is introduced and explore it in context. Understanding often comes from seeing how ideas are used, not just how they are defined.

About the Authors

Antonios E. Marsellos is a professor in the Department of Geology, Environment, and Sustainability at Hofstra University, where his work focuses on using computers to study Earth data (geoinformatics), Earth data analytics, and environmental modeling. His interdisciplinary research integrates spatial analysis, time series methods, and data-driven approaches to better understand natural hazards, climate processes, and Earth system dynamics.

He has taught geostatistics, spatial analysis (GIS), and data analytics across universities in Europe and the United States, with an emphasis on hands-on learning and research-based education. His work with students has demonstrated that complex analytical concepts can be made accessible through coding, enabling learners with no prior experience to develop research projects and present their work at scientific conferences and competitions.

His expertise has been featured in national and international media, including NBC News, ABC, FOX 5, and Deutsche Welle, where he has contributed to public discussions on earthquakes, volcanic activity, and environmental hazards. Through both research and teaching, he is committed to making data analytics a practical and accessible tool for understanding Earth systems.

Katerina G. Tsakiri is a professor in the Department of Information Systems, Analytics and Supply Chain Management at Rider University. Her work focuses on data analytics, decision-making systems, and the application of quantitative methods in business and environmental contexts.

Her approach emphasizes clarity, interpretation, and the practical use of analytical tools for real-world problem solving. She brings a complementary perspective to this book by connecting data science principles with broader applications in analytics and decision-making.

Together, the authors share a common goal: to make Earth Data Analytics accessible to a wide audience and to support students at all levels in developing the skills needed for data-driven research. They are also co-founders of ResForMe (resforme.com), an educational initiative focused on mentoring high school students in research and data analytics.

www.ingramcontent.com/pod-product-compliance
Ingram Content Group UK Ltd.
Pitfield, Milton Keynes, MK11 3LW, UK
UKHW051959041026
15590UKWH00009B/2378

9 798996 563500